ELECTRICAL INSTALLATION COMPETENCES

Part 2 Studies: Practical

Maurice Lewis
BEd (Hons), FIEIE

Stanley Thornes (Publishers) Ltd

© Maurice Lewis, 1994

Original line illustrations © Stanley Thornes (Publishers) Ltd 1994

All rights reserved. No part of this publication may be reproduced or transmitted in any form or by any means, electronic or mechanical, including photocopy, recording or any information storage and retrieval system, without permission in writing from the publisher or under licence from the Copyright Licensing Agency Limited. Further details of such licences (for reprographic reproduction) may be obtained from the Copyright Licensing Agency Limited, of 90 Tottenham Court Road, London W1P 9HE.

First published in 1994 by:
Stanley Thornes (Publishers) Ltd
Ellenborough House
Wellington Street
CHELTENHAM GL50 1YD
United Kingdom

A catalogue record for this book is available from the British Library.

ISBN 0 7487 1659 9

Cover photograph by courtesy of
MEGGER INSTRUMENTS LIMITED

Typeset by Florencetype, Kewstoke, Avon
Printed and bound in Great Britain by The Bath Press, Avon

Contents

Preface v

1 Safety at work 1

Introduction; safety awareness; examples 1–4; statutory requirements; exercise 1.

2 Installation practices and technology 8

Isolation of supplies; using a clamp-on ammeter; voltage measurement in 3-phase system; voltage drop in an electrical installation; instrument connections; residual current devices; selection of a wiring system; dismantling and reassembling a three-phase motor; current measurement in a radial circuit; 3-phase induction motor test; a fluorescent lamp circuit; operation of a contactor; lighting circuits; current measurement in a SON lamp; voltage measurement in test instruments; testing an electrical installation; testing electronic equipment; measurement of amplitude and frequency; fault finding in a ring circuit and lighting circuit

3 Course assignments 42

Introduction; installation drawings; engineering workshop; two-bedroom flat; motor installation; staff rooms in a private school; college laboratory workshop; hotel restaurant; conduit and trunking exercises; TN-C-S earthing system; staff dining room; luxury house.

4 Laboratory work 61

Guidance notes for laboratory work; star-delta connections; RC in series; RL in series; RLC in series; RLC components in series and parallel; SOX discharge lamp; shunts and multipliers; frequency variation; universal motor speed–torque characteristics; voltage drop in a cable; transformer ratios; temperature coefficient of resistance; three-phase induction motor; power factor correction; rectifier circuits; audio frequency amplifier; multimeters; earth fault loop impedance; star-delta starter; transformer tests.

5 Terminology and further information 88

Preface

Part 2 Studies: Practical is the fifth book in the electrical competences series, covering practical topics in the City and Guilds 2360–8 scheme. The book is specifically written to cover tasks and performance criteria required in centre assessment component 2360–8–24.

Chapter 1 is an introductory chapter on safety at work, describing to you some of the author's own experiences when he was a practising electrician. It outlines some of the important responsibilities employers and employees have under the Health and Safety at Work Act and other statutory regulations.

Chapter 2 covers the tasks found in college assessment record books and provides you with many easy-to-follow exercises to complete. Your college or training staff may find these exercises beneficial and adapt them to their own teaching programme.

Chapter 3 is concerned with course assignments, focusing attention on installation drawings. Here, you have the opportunity to draw block and circuit diagrams and prepare material requisitions from layout drawings. In this chapter various exercises encourage you to participate and look for further information elsewhere.

Chapter 4 covers the essential technology of the course. Most of the mystery surrounding science-related topics can be understood by doing laboratory work. This chapter hopes to strengthen your written communication, asking you to complete a number of set experiments.

Chapter 5 is concerned with terminology and reference material and introduces you to all the terms used in the other chapters item by item. Each item contains a lot of useful additional information.

Maurice Lewis

Acknowledgements

Throughout this book the author has made reference to the following organisations and their publications:

British Standards Institution – BS 3939 Graphical symbols for electrical power, telecommunications and electronic diagrams.

City and Guilds of London Institute – Course 2360 Electrical Installation Competences.

Health and Safety Executive – Guidance Note GS 38

Institution of Electrical Engineers – Regulations for Electrical Installations, Guidance Notes and Site Guide.

The author thanks these organisations and hopes that students will benefit from the further reading.

About the author

Maurice Lewis is currently an independent safety consultant for electrical contractors and industrialists in the East Anglia region. He specializes in the 16th Edition of the IEE Wiring Regulations and the Electricity at Work Regulations.

He spent 27 years in teaching and became Head of the Electrical Section at Luton College HE in 1986. As an ongoing interest, Maurice works voluntarily with City and Guilds of London Institute as a visiting assessor for the Association of Colleges in the Eastern Region.

Safety at work

Objectives

After working through this chapter you should be able to state a number of duties imposed on employers and employees relevant to the following statutory regulations:

- 'Health and Safety at Work Act 1974'
- 'Construction (Head Protection) Regulations 1989'
- 'Electricity at Work Regulations 1989'

With supporting references, you should be able to:

- describe the safety procedures when using access equipment;
- know the requirements for electrical supplies on construction sites;
- know the procedure for completing a permit-to-work document;
- know a number of safety precautions when terminating mineral insulated cable;
- describe the operation of an RCD and state the IEE Regulations requirements for its use on TN and TT earthing systems;
- describe the dangers when disposing discharge lamps.

1.1 Introduction

When I started my electrical apprenticeship in the early fifties, the medium size firm that I worked for had an interest in both electrical engineering and electrical installation contracting. The engineering business dealt with shipping and dockyard work while the contracting business dealt with a range of projects in both commercial and industrial premises.

I spent most of my first year in the stores and occasionally in the office, answering telephone calls during lunch break. The instructions were quite simple, 'Be polite, take down the person's name and telephone number and say that I (the managing director) would be back soon.'

Like most apprentices, I had to attend college to obtain the City & Guilds 'B' (Part 2) Certificate. The IEE Regulations in force at the time was the 12th Edition, superseded on the 1 September 1955 by the 13th Edition. If you counted the eleven regulations in Part I 'Requirements for Safety', you would find it contained 143 regulations and all these had to be remembered for exam purposes. I used to like the definition of a final circuit (called a final sub-circuit) which was: 'An outgoing circuit connected to a distribution board and intended to supply electrical energy direct to current-using apparatus.' The voltage bands covered extra-low, low, medium and high voltage. Voltage drop from a consumer's terminals to any point in the installation had not to exceed 1 volt plus 2% of the declared voltage, with some exemptions.

In the stores, I obtained a good understanding of electrical products as well as various types of wiring system, different tools and access equipment. Our storeman was very strict and would not allow anything to be thrown away, neither a nail nor a woodscrew, not even a bent buckle clip. Every item that was returned from a completed job was noted and every length of unused cable measured. Portable drills and other electrical equipment were always tested before being let out again.

I found the stores discipline of immense value in the years that followed; it made me always keep a tidy workplace in order to take stock of what I was doing. It also gave me a good understanding of ordering materials from catalogues as well as a sound knowledge of test equipment. Towards the end of my first year, I was sent to the firm's main shipping office in Cardiff docks to gain some experience on ships. I remember well my first day on board an oil tanker, when instructed to get in the boatswain's chair and fit a lamp at the top of the mast. The swinging experience wasn't too bad, but it was quite frightening looking down and I don't think I could ever manage such a task out at sea.

It is not until you are working on a construction site or on some major project that life as an apprentice electrician takes on new meaning. You soon realise how variable your work can be and you meet people in your own firm who have special expertise in some area of installation work. You also meet people in other trades and see how they organise and operate their work schedules. You soon realise that your work, while important, is only part of a general plan, to complete a contract in some allotted time. It is often the case in industry, that when a trade contractor falls behind his contract time that 'corners are cut' and sadly, safety procedures can get overlooked.

1.2 Safety awareness

Every year, about 500 people die and several hundred thousand people suffer illness or injury caused by accidents in the workplace.

In those days the *Health and Safety at Work Act* did not exist but an awareness of safety rules was brought home to me rather abruptly. It happened during the second year of my apprenticeship while working in the steel works at Ebbw Vale. Each year, the plant would make a huge profit selling its steel, yet despite this achievement, its accident record was far from satisfactory. The following anecdotes should convey to you the importance of clear communication and proper procedure if accidents are to be avoided. After reading these examples, you might still wonder what else can be done.

Example 1

One day, we had to erect a triple extension ladder inside a loading bay in order to inspect the wiring to a lighting point. I was asked to help the foreman and electrician's mate carry the ladder since it was long, made of wood and very heavy. At the place of working and to get the best climbing angle, the ladder was erected in the middle of a railway line. It had to reach up to a steel catwalk some 25 m high. Not too far away from us on the railway line stood eight empty trucks and thinking of safety before work commenced, the foreman notified the Work's railway depot. They gave permission and issued the foreman with a red danger flag which he had to attach to the last truck.

The foreman climbed the ladder first and lashed the top to the catwalk. To assist him with his inspection work the electrician's mate also climbed the ladder. While only halfway up the trucks on the line, quite unexpectedly, started shunting together. I saw that they had been hit by a free running truck and the collision caused the last truck to come hurtling down the siding towards us. Frantically, I shouted to the mate to hurry up climbing and was forced to take drastic action myself by running for cover. Needless to say, the truck completely destroyed the ladder with pieces flying everywhere and it left the electrician's mate clinging on to the wrists of the foreman who eventually rescued him.

While no one received injury, in hindsight, the foreman should not have accepted a verbal reply from the railway depot that everything would be all right. The red danger flag was insufficient. It was discovered afterwards that there was lack of communication between two groups of shift workers and the train driver responsible for the shunting duties was not informed of the work being carried out.

Example 2

One job we were asked to do was to install special gas-tight, galvanised conduit to protect supply cables feeding an electric furnace. The job was in an extremely hostile environment, very noisy, very dirty and highly dangerous. We were compelled to wear safety helmets and use barrier cream to protect our hands. The specially-made 3 in diameter conduits were quite heavy and came in 2 m lengths. They had to be threaded on a lathe in the Work's tool room. One day, two apprentices were instructed to take a length of pipe to the tool room for it to be cut and threaded. On their way there the pipe was accidentally dropped and it badly injured the foot of one of the apprentices. Afterwards, it came to light that the two lads had been 'messing about' and had not adopted a planned procedure for carrying the pipe. The injury might not have been so severe if the apprentice had worn protective footwear.

It was on this same job that a welder and his mate were fatally injured when their gas cylinder exploded and in the very same week a plant operator died after falling into a container of molten metal. It was thought that he had fainted from the excessive heat while raking off slag from a ladel.

Example 3

Towards the end of my five-year apprenticeship (the approval stage), I was given the task of wiring the high bay lighting in the Work's hot-strip mill. This was my first major job and it was difficult because the work had to be done from off the top of a gantry crane. Like the previous example, the environment was hostile, very noisy and dirty with noxious fumes drifting into the atmosphere from the continuous action of the Work's mechanical stands. The purpose of these stands was to reduce the size of the rectangular shaped iron ingots, at one end of the mill, to wafer thin metal at the other end of the Mill, where it would be coiled and cooled.

The crane's controls had to be modified so that it could be operated at ground level. My first task was to design a prototype angle-iron bracket that could be fixed easily to the girder and also be strong enough to take the weight of the lighting fitting. Once this was done, it was then a question of how long my team of helpers (one apprentice, one electrician's mate, a welder and his mate) would take to make more brackets.

Over several weeks the job was going extremely well and I would spend most of my time fixing the brackets on the girders at the required distances. One day, when I was sitting on the traverse of the crane, it started to move forward and I was knocked backwards, falling off the traverse onto the crane's platform. If I had taken up my usual position on the traverse, I would have fallen through the middle of the crane and been killed.

The accident was caused by an operative in the mill taking hold of the controls and wanting to use the crane for his own purpose. Apparently, he had returned from holiday and was not told by the maintenance department that the crane had been allocated to the electrical contractors. The annoying thing, however, was that a danger notice was placed over the crane's controls with words saying 'men at work'. Even more annoying to me was the fact that my apprentice had left the controls unattended.

Example 4

As a final example, I had the job of running a PVC armoured cable in the Work's tinplate mill. The cable supplied a large three-phase compressor motor and was fed from a relay room. The cable route was long and through all kinds of obstruction. Several types of installation method were used, including ducting, traywork, clipping to walls and fastening to girders.

At the connection stage, I allowed an electrician to terminate the cable in the motor and two apprentices to terminate the cable in the relay room. The supply in the control panel was switched off but not in the adjacent panels. All the panels were without rear doors to close and it was necessary to erect protective insulated screens to avoid possible danger.

The apprentices were told to allow adequate spare cable during the terminating process so that it could be neatly clipped inside the panel. They were also told not to unravel too much free armouring but to bend it in half when removing it from the cable. During my checking of the work and inspection of the motor connections I returned to the relay room to discover that a massive short circuit had occurred. The apprentices had not listened to my instructions and in the process of stripping the cable's armouring, managed, quite unbelievably, to find contact with live connections in one of the adjacent panels. The accident was put down to carelessness and a certain amount of tomfoolery from the apprentices. Both they and the electrician were very fortunate not to receive electric shock.

In summary, it is hoped that these few examples have illustrated to you the dangers associated with electricity and how careful you must be when communicating your wishes to others to achieve a safe system of work. It is not enough to obtain verbal or written instruction before work is carried out: you must continually monitor the system and review it in the light of changed circumstances.

1.3 Statutory requirements

Over the last 10–20 years, people working in the construction industry (including the electrical contracting industry) have been constantly informed on matters of general safety.

This awareness is mainly attributable to the *Health and Safety at Work Act 1974* with its statutory powers of ensuring that employers and organisations have a duty to protect you and keep you informed about health and safety matters.

Under the *HSAW Act* your employer has a number of responsibilities, some of which are to:

- assess your workplace to see what hazards are present;
- take steps to remove or control any possible risks by keeping dust, fumes and noise under control;
- make sure that any protective measure taken is effective;
- provide adequate welfare facilities;
- draw up a health and safety policy if there are five or more employees;
- provide free, any protective clothing or equipment;
- provide adequate first aid facilities and report certain injuries, diseases and dangerous occurrences;
- set up a safety committee if asked in writing by two or more safety representatives;
- provide you with information, instruction and training.

As an employee, you have a responsibility to:

- co-operate with your employer on health and safety matters, such as wearing personal protective equipment (e.g. safety helmet, gloves, industrial footwear, eye protection, etc);
- take reasonable care of your own health and safety and that of others who may be affected by what you do (or don't do);
- not interfere with or misuse anything provided for your health and safety at work.

One long-awaited piece of HSE legislation which came into force on the 30th March 1990 was the *Construction (Head Protection) Regulations 1989* which required suitable head protection to be worn during construction work, unless there was no risk of injury from falling objects or a person hitting his/her head against something on site. In the Regulations' first year, it was estimated 250 construction workers were saved from having head injuries.

Another fairly recent piece of legislation for electrical dutyholders is the *Electricity at Work Regulations 1989*. These regulations replace the older *Electricity (Factories Act) Regulations* of 1908 and 1944 and bring a lot more premises into line with the present law. It places duties on all people concerned with electrical work and discourages live working.

The first task that you have to do in Chapter 2 is concerned with isolation of supplies. This is a requirement of Reg 12 of the *Electricity at Work Regulations* and is extremely important. The device used for this function must meet the following requirements:

- not allow inadvertent reconnection;
- establish a positive break from the live conductors;
- establish adequate creepage clearance;
- provide a positive indication of its open and closed position;
- incorporate means to prevent unauthorised operation (such as a means of locking off);
- only be common to several items of equipment where the equipment is normally operated as a group.

By providing clear identification and labelling of circuits, protective devices and switchgear, you will minimise the risk of incorrect isolation. It is important to remember that main switchgear needs to be sited so that it is accessible for operation as well as maintenance and it must also be made secure against unauthorised interference.

For further information on other related statutory regulations, see Chapter 1 'Requirements for safe working practice' in the author's *Part 1 Studies: Theory* book of this series.

In the following chapters the emphasis is placed on you to carry out installation and laboratory tasks in a safe manner. Make sure that all your electrical connections are sound and if using a tester or measuring instrument, that you understand how to use the instrument properly. Quite a number of accidents are caused by distraction or by not following simple instructions.

EXERCISE 1

1. With reference to Figure 1.1, describe several ways in which the accident through over stretching on the ladder could have been avoided.

Figure 1.1 The consequences of over stretching when working on a ladder

2. a) Distinguish between those regulations which are classified as 'statutory' and those which are classified 'non-statutory'.
   b) What is the legal status of the 16th Edition of the IEE Wiring Regulations now that it is a British Standard?
3. On a construction site, electrical supplies are often through BS 4363 equipment and BS 4343 socket outlets, Draw a line diagram of a single-phase distribution system providing 110 V to socket outlet units.
4. Figure 1.2 shows a number of hazards found on a construction site. Make a list of at least ten, giving a brief description of how and why they should be avoided.
5. a) What is a permit-to-work document?
   b) List FOUR precautions that can be taken when working on a live installation to avoid electric shock.
6. A three-phase, direct-on-line contactor starter is to be dismantled for maintenance. Make a list of the sequence that should be adopted which should include commissioning the starter.
7. Figure 1.3 shows a diagram of mineral insulated cable termination.
   a) Label the components marked A to I
   b) Write a step-by-step procedure for terminating the cable and state at least FIVE safety considerations.
8. a) With the aid of a circuit diagram, explain the operation of a residual current device.
   b) State the IEE Wiring Regulations for its use in TN and TT earthing systems.
9. When the operating life of discharge lamps is reached, the lamps should be disposed of by a mechanical crushing machine. Describe some of the dangers that might result if such a crushing machine is not used and the lamps were broken on site by hand.
10. Mobile scaffold towers are covered by *The Construction (Working Places) Regulations 1966* and in HSE Guidance Note GS 42. State a number of requirements for siting, erecting, moving and climbing the tower.

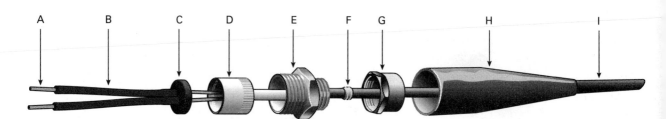

Figure 1.3 MI cable termination

Figure 1.2 Hazards often found on a construction site

Installation practices and technology

Objectives

After working through this chapter you should be able to:

- *isolate the supply of electricity in an electrical installation;*
- *use a 'clamp-on' ammeter to measure current in an unbroken circuit;*
- *measure voltage drop in an electrical installation;*
- *measure voltages in 3-phase, 3-wire and 4-wire circuits;*
- *measure power, voltage and current in 3-phase and 1-phase circuits;*
- *measure the tripping time of residual current devices;*
- *install a wiring system to Regulation requirements;*
- *measure current, terminal voltage and voltage drop in a radial circuit;*
- *connect a 3-phase cage induction motor to a direct-on-line starter;*
- *dismantle and reassemble a three-phase induction motor;*
- *install and operate a triple-pole contactor through open and closed contacts;*
- *wire and connect various types of lighting circuit;*
- *measure current in a SON discharge lamp circuit;*
- *measure voltage in different parts of a fluorescent lamp circuit;*
- *carry out prescribed tests on an electrical installation;*
- *carry out tests on electronic components using electronic instruments;*
- *measure the input and output waveforms of an audio amplifier;*
- *diagnose faults in a 30 A ring circuit;*
- *diagnose faults in a SON lamp circuit.*

2.1 Isolation of supplies

Task aims

To show how to isolate safely both an electrical system and a final circuit, and carry out relevant confirmatory tests.

Task objectives

You should:

1. identify items of electrical equipment in an installation which are regarded as isolators (see the *IEE Wiring Regulations*, Guidance Note No. 2, Sects. 2 & 3);
2. show the safe isolation of an electrical supply to an installation at (a) the origin and (b) a final circuit;
3. perform confirmatory tests to ensure that the electrical installation is 'dead' and safe to work on.

Observed competence

The person supervising your practical work will be looking for the following points:

1. Your understanding of the term 'isolation' and the purpose of isolation.
2. The names of devices used to perform the function of isolation.
3. The correct procedure in securing isolation of the whole electrical installation and a final circuit.
4. Your understanding of the confirmatory tests to be made on the installation to ensure that it is 'dead' and remains 'dead' while work is being carried out on it.

Guidance notes

1. Regulation 12 of the *Electricity at Work Regulations 1989* states that suitable means be made available for (a) cutting off the supply of electrical energy to any electrical equipment and (b) the isolation of any electrical equipment. These two functions, often referred to as **disconnection** and **separation**, are not the same since the latter requires the electrical equipment to remain switched off so that it cannot be accidentally reconnected.

2. The purpose of isolation is to enable a skilled person(s) to carry out work on, or near, normally live parts, and so avoid the dangers caused by electricity.

3. All isolating devices, such as isolating switches, circuit breakers, plugs and socket-outlets, fuselinks and links, switch-fuses and fused switches, must be durably marked in order to identify the installation or circuits they control. They should be placed in suitable positions and be easily accessible. To prevent unintentional reclosure the isolating device should be lockable. Some devices such as a lockable miniature circuit breakers are acceptable provided their contacts establish an adequate opening distance of 3 mm (for devices operating on 240 V a.c.). Figure 2.1 shows some examples of isolating devices.

4. Before you isolate an electrical installation or circuit you should:

    a) be a competent person;
    b) inform others of your intention;
    c) follow a safe system of work.

    A competent person is described in Regulation 16 of the *Electricity at Work Regulations 1989* and also on page 34 of the Regulation's *Memorandum of Guidance*.

5. There is always some reason for isolating the whole or part of an electrical installation, whether it be for maintenance, fault finding, alterations or additions to the installation. You should try to obtain as much information as possible from others about the work you intend doing. Stay within the law by keeping yourself up-to-date with relevant HSE leaflets and codes of practice. You should adopt a safe system of work approach and know your company's safety policy. For work on high voltage systems or complicated systems the *Joint Industry Board's Safety Bulletin No. 6 (Revised)* should be read as it covers aspects of a **Permit to Work Document**. More information on the procedures for isolation can be found in Guidance Note 2 of the *IEE Wiring Regulations* and also Regulation 13 of the *Electricity at Work Regulations* concerning precautions for work on equipment made 'dead' (see also page 28 of the *Memorandum of Guidance*).

6. Isolation of the whole electrical installation is best achieved when the premises are unoccupied, such as, before or after working hours or

Isolation

Cutting off an electrical installation, a circuit, or an item of equipment from every source of electrical energy.

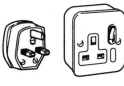
Plug and socket outlet

Isolator

Switch-fuse

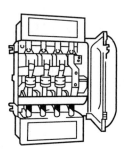
Fused switch

Devices suitable for isolators

- Isolators (disconnectors)
- Isolating switches (switch disconnectors)
- Plugs and socket outlets
- Fuse links
- Links
- Circuit breakers having the required contact separation

Lockable D.O.L. switch

Residual current device

Fuseboard incorporating a D.P. switch

Where isolation is required

(i) At the origin of every installation
(ii) For every circuit or group of circuits
(iii) For every motor and associated control gear
(iv) For every discharge lighting circuit exceeding low voltage
(v) For maintenance of main switchgear
(vi) If remote from the equipment

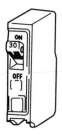
Lockable M.C.B

Removable link

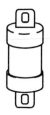
Fuse link

Figure 2.1 Electrical equipment used for isolating puposes

during a maintenance shut-down period. The owner, occupier or person(s) responsible for the building (e.g. a building officer or work's manager) should be notified.

7 The following steps can be taken to provide a safe system of work:

(i) securing the means of isolation in the open position;
(ii) posting warning notices;
(iii) issuing safety documentation, where appropriate.

8 Isolation of a final circuit from, say, a distribution board will of course depend upon the degree of danger present. Some distribution boards and their circuit protective devices can be isolated and locked off, but those that cannot be locked off require additional safety precautions to be taken, such as removing the protective device and disconnecting circuit conductors. It is important to prove the circuit is 'dead' using a reliable voltage tester. This should be done before and after work commences on the circuit. To inform others that the circuit is isolated, you should post a warning notice on the distribution board.

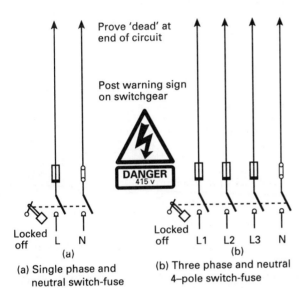

Figure 2.2 Isolation of supplies

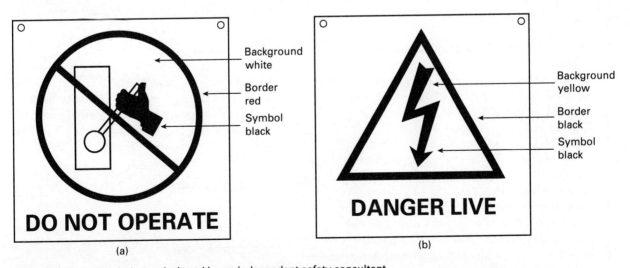

Note: These two symbols are designed by an independent safety consultant

Figure 2.3 (a) 'Do not operate' notice, for attachment at points of isolation (prohibition format)
(b) 'Danger' notice, for attachment to adjacent live apparatus (warning format)

2.2 Using a clamp-on ammeter

Task aims

To measure the current in a three-phase, three-wire and four-wire system using a **clamp-on ammeter** similar to Figure 2.4.

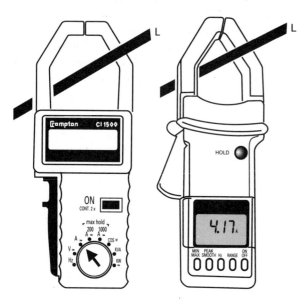

Figure 2.4 Typical clamp-on current meters with facilities to cover ranges of current a.c. or d.c. up to 1000 A

Task objectives

You should:

1 study Figure 2.5 and check the current ratings of the chosen cables against the rated currents of the connected loads (see the relevant cable tables of current carrying capacity and voltage drop in Appendix 4 of the *IEE Wiring Regulations*);
2 state the important regulation in Chapter 13, Part 1 of the *IEE Wiring Regulations*, concerning the size and current-carrying capacity of circuit conductors;
3 inspect a proprietary clamp-on ammeter and read the instructions on how to use the instrument, then select the correct scale range;
4 switch on the supply and apply the clamp-on ammeter to each load (see Figure 2.5). Measure the current in each phase conductor and the neutral conductor of the heating load. Create an imbalance between the phases of the heating load and take further readings of the current in each circuit conductor.
5 record all your results and state TWO advantages of using the clamp-on ammeter over a standard ammeter connected in circuit.

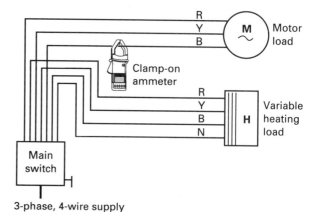

Figure 2.5 Current measurement using a clamp-on ammeter

Observed competences

The person supervising your practical work will be looking for the following points:

1 your understanding of cable sizes and their current-carrying capacities;
2 your understanding of the clamp-on ammeter and your ability to use it in three-phase circuits;
3 your ability to record results and explain the advantages of using a clamp-on ammeter.

Guidance notes

1 Study Figure 2.5 and check the sizes of the cables used for each circuit. You may have to consult a cable manufacturer's catalogue or an electrical wholesaler catalogue in order to find the cross sectional areas of single-core insulated flexible cords. As a guide:

   0.5 mm^2 is rated at 3 A;
   0.75 mm^2 is rated at 6 A;
   1.0 mm^2 is rated at 10 A;
   1.25 mm^2 is rated at 13 A;
   1.5 mm^2 is rated at 16 A.

2 Obtain a catalogue of test equipment and read the details of its clamp-on ammeters both analogue and digital.

3 Study the design of a typical clamp-on ammeter and read its operating instructions for measuring current.
4 Select a suitable current range for the ammeter, then switch on the supply to the loads and record the current in each phase and neutral conductor as described above.
5 With the exception of switchboard ammeters, it is not normal practice to monitor the current flow in every sub-main cable or final circuit. However, occasions arise in large commercial and industrial premises to find the circuit phase and neutral current. The inclusion of a standard ammeter in circuit conductors already connected is sometimes impossible to achieve, apart from not knowing the size of ammeter required, particularly if the load current is fluctuating. The advantage of using a clamp-on is plain to see with its ease of use and multi-purpose use on d.c. supplies as well as for measuring other quantities such as voltage, resistance and frequency.

2.3 Voltage measurement in 3-phase systems

Task aims

To measure the voltage in three-phase, three-wire and three-phase, four-wire circuits using a proprietary voltage tester or a multimeter type instrument selected for voltage measurement.

Task objectives

You should:

1. inspect the chosen instrument for its use as a voltage tester and read the instructions on how it is used;
2. inspect the instrument's test leads for any signs of damage;
3. select a suitable voltage range on the tester so that it will read the supply voltage and circuit voltages;
4. study Figure 2.6 and measure the line and phase voltages between all the circuit conductors;
5. record results and make comments on the differences found.

Observed competences

The person supervising your practical work will be looking for the following points:

1. Your understanding of the voltage tester and its use on different voltage ranges.
2. Your understanding of the dangers present when using the tester.
3. Your understanding of HSE *Guidance Note GS38*.
4. Your ability to measure accurately and record voltages in three-phase circuits.

Guidance notes

1. Obtain an appropriate instrument for making voltage tests and read its instruction leaflet.
2. Select a voltage range on the tester that will be appropriate for the a.c. circuit under test.
3. Securely connect the test leads to the tester and then switch on the supply.
4. Carefully apply the test leads to the terminals under test, making a positive contact.
5. Record the voltages on the tester between each pair of phase terminals (e.g. R–Y, R–B, Y–B and also R–N, Y–N and B–N).
   Note: for more accuracy, you should change the voltage range when measuring phase to neutral voltages.
6. It is important to check to see if the tester is selected to read voltage and not another quantity as this might cause damage to the tester or its leads. Sparking is a possibility between the test leads and the terminals under test. Some test instruments are fitted with a safety fuse or they may have a cut-out which can be re-set. Test leads designed to HSE *Guidance Note GS38* incorporate fuse protection and are provided with lavish insulation. Only a bare minimum of metal is exposed when the probes are in contact with live terminals.

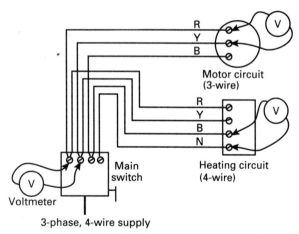

Figure 2.6 Measuring in three-phase, three-wire and three-phase, four-wire circuits

2.4 Voltage drop in an electrical installation

Task aims

To compare voltages in various parts of an electrical installation and to find the percentage voltage drop.

Task objectives

You should:

1. study Figure 2.7 and take voltage readings at the places indicated;
2. record voltage measurements and determine the percentage voltage drop;
3. write comments on your findings.

Observed competences

The person supervising your practical work will be looking for the following points:

1. Your ability to use a voltmeter and take voltage readings in different parts of the electrical installation.
2. Your ability to determine the percentage voltage drop.
3. Your understanding of Section 525 of the *IEE Wiring Regulations* concerning voltage drop in consumers' installations.

Guidance notes

1. Obtain an appropriate instrument for making voltage tests and read its instruction leaflet.
2. Select a voltage range on the tester so that it

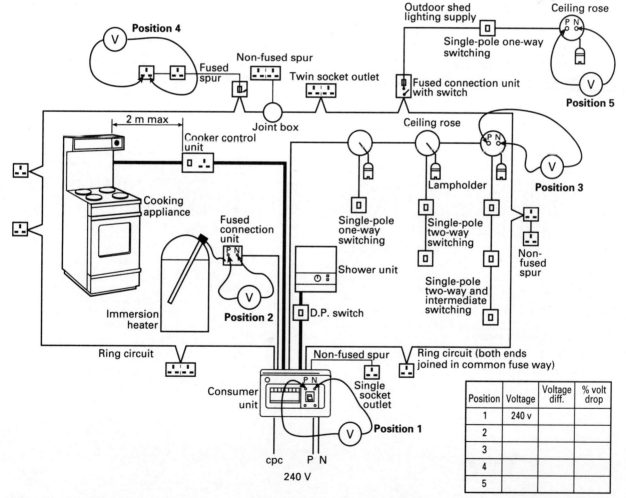

Figure 2.7 Measuring voltage in different parts of an electrical installation

15

will be appropriate for the a.c. supply voltage and the final circuits under test.

3 Securely connect the test leads to the tester and then switch on the supply to the installation.
4 Carefully apply the tests leads to the terminals under test, working through the installation as indicated in Figure 2.7.
5 Record the voltages on the tester and determine the percentage voltage drop. Maximum voltage drop (V_{max}) allowed by Reg 525–01–02 is 4% of the supply voltage. For the installation supplied at 240 V, V_{max} = 4/100 × 240 V = 9.6 V. This maximum voltage drop is from the intake position to the furthest point in the installation. To record your percentage voltage drops subtract the circuit voltages from the supply voltage. The percentage voltage drop = (voltage difference × 100)/240 V. For example, say, at position 4 the voltage was 210 V. The voltage difference is 240 V – 210 V = 30 V. Therefore, X% = (30 × 100)/240 = 12.5%

2.5 Instrument connections

Task aims

To measure the power, voltage and current in single-phase and three-phase a.c. circuits.

Task objectives

You should:

1. connect the circuit instruments as shown in Figure 2.8 and take measurement of power, voltage and current;
2. record your results and calculate the circuit voltamperes (VA) from the ammeter and voltmeter readings;
3. compare the VA readings with the wattmeter readings.

Guidance notes

1. Connect the instruments as shown in Figure 2.8 (a), making sure the wiring is correct to the wattmeter.
2. Switch on the single-phase a.c. supply and record the instrument readings.
3. Repeat the above procedure for the three-phase circuits shown in Figures 2.8 (b) and 2.8 (c).
4. Determine the voltamperes for each circuit and tabulate your results. Note that in a three-phase circuit the power consumed is expressed as

$$P = \sqrt{3}\, V_L I \cos\phi$$

Since resistors are used in this circuit, $\cos\phi = 1$.

5. Compare the wattmeter reading with the voltamperes. You might find small errors due to the instruments and the connection leads.
6. As an additional exercise, incorporate an inductor in each circuit and repeat the above procedure. Remember that the power factor ($\cos\phi$) will be less than unity and the voltamperes

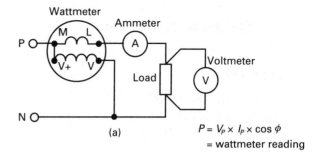

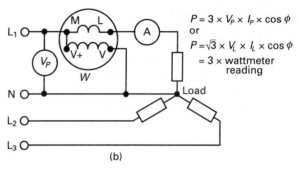

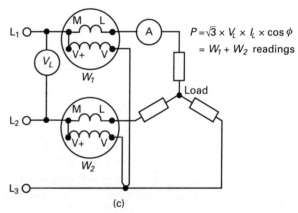

Figure 2.8 Instrument connections (a) single-phase; (b) three-phase, four-wire; (c) three-phase, three-wire

(apparent power) will exceed the wattmeter reading (true power). The ratio of these two quantities gives the power factor of the circuit, i.e.

$$\text{p.f.} = \frac{P}{(\sqrt{3}\, V_L I)}$$

2.6 Residual current devices

Task aims

To measure the tripping time of three residual current devices (RCDs) and to compare their operation with test requirements.

Task objectives

You should:

1. read manufacturers' catalogues on the specification and operation of RCDs;
2. make reference to relevant literature concerning the requirements for RCDs;
3. make time checks on different RCDs to confirm their operation according to their tripping sensitivities;
4. tabulate test results and write comments on the findings.

Observed competences

The person supervising your practical work will be looking for the following points:

1. Your understanding of the operation and testing procedure for RCDs.
2. Your understanding of the regulation requirements for testing RCDs.
3. Your ability to perform and record tests on RCDs with the aid of a proprietary test instrument.

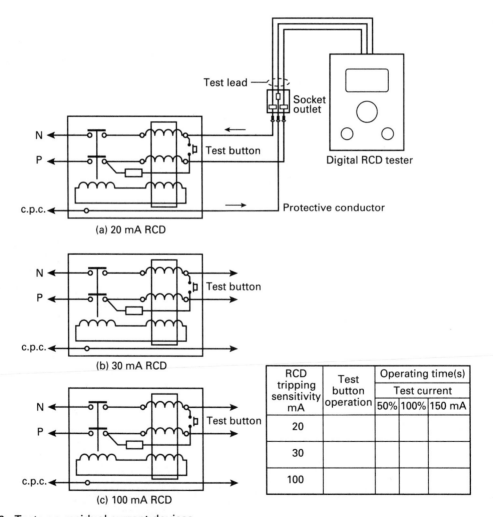

Figure 2.9 Tests on residual current devices

RCD tripping sensitivity mA	Test button operation	Operating time(s) Test current		
		50%	100%	150 mA
20				
30				
100				

Guidance notes

1. For information on the specification and operation of RCDs, consult a local electrical wholesaler or write directly to a product manufacturer. Journals such as *Electrical Contractor* and *Electrical Installation* often publish articles on RCDs and you will also find considerable information in the *IEE Site Guide* and the *IEE Wiring Regulations* and *Guidance Notes No. 1* (page 15) and *No. 3* (page 81).

2. Study Figure 2.9 and connect the 20 mA RCD in circuit via the 13 A plug. Now carry out the following tests:

   a) Switch on the RCD and check its operation with the test button.
   b) Read the operating instructions of a proprietary RCD tester and then insert its plug into the RCD socket outlet.
   c) Select a tripping current of 10 mA (50%) and press the tester's operating button for 2 s. The RCD should not operate.
   d) Change the current setting on the tester to 20 mA (100%) and again press the operating button.
   e) Record the time taken for the RCD to trip.
   f) Repeat step (e) above but with a setting of 150 mA and again record the time for the RCD to trip.
   Note: the maximum time for making the test at 150 mA is 50 ms.

3. Repeat steps (a) to (f) for the 30 mA and 100 mA RCDs and tabulate the results of each test.

4. The *IEE Wiring Regulations* require the test button on an RCD to be tested quarterly by the user (see Regulation 514–12–02). This is to check its mechanism and also its sensitivity. The test at 150 mA is to show that the RCD will trip in a time of 40 ms. This is the maximum time allowed for a person to be in contact with 240 V a.c.

2.7 Selection of a wiring system

Task aims

To install a suitable wiring system in a domestic dwelling to supply a 3 kW/240V electric heater. The following assumptions are made:

- the heater is installed as a fixed appliance;
- the supply to the heater will be from a spareway in the consumer unit;
- the circuit protective devices in the consumer unit are BS 1361 cartridge fuses;
- the length of run is 15m;
- thermal insulation affects one side of the cable;
- the external earth loop impedance is to be taken as 0.35 Ω.

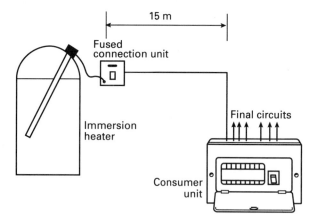

Figure 2.10 Immersion heater circuit

Task objectives

You should:

1. Select a suitable wiring system from one of the following:

   a) single-core PVC cables drawn into metal conduit;

   b) two-core mineral insulated metal sheath cable;

   c) two-core, PVC-insulated and sheathed cable incorporating cpc.

2. Make reference to the building construction, environmental conditions, load current and method of overcurrent protection.

3. Select and install a suitable size cable for the chosen wiring system.

Observed competences

The person supervising your practical work will be looking for the following points:

1. Your reason for choosing the selected wiring system.
2. Your understanding of Appendix 4 of the *IEE Wiring Regulations*.
3. Evidence of the steps taken to select a suitable cable satisfying the above circuit conditions.

Guidance notes

1. You should make your selection based on usual practice or a requirement of an electrical specifications. One important consideration is the cost of the wiring system in terms of installation time and materials. Metal conduit systems, while providing good mechanical protection are expensive to install. They lack the flexibility (ease of wiring) offered by the other two systems. MIMS cables and accessories are expensive and their termination procedure increases installation time. PVC-insulated and sheathed cables are not so expensive as the other two wiring system. They are flexible and provide a simple termination procedure.

2. It is usually the practice in most domestic dwellings of brick and timber construction to install a flush wiring system using twin and cpc flat cables to BS 6004. As a general rule, few environmental conditions apply. For example, the ambient temperature is not likely to be a problem except where the cables terminate in enclosures, or where heat is generated (in luminaires and in heating appliances). For the maximum conductor operating temperature of general-purpose PVC, see the *IEE Wiring Regulations*, Reg 523–01–01 and Table 52B. Where the cable is run in thermal insulation, reference should be made to Reg 523–04–01. Grouping factors are not normally applied to PVC cables used in domestic premises. From the point of view of load current, it is usually the practice to change the load current into design current. This will then take into consideration the conditions within the installation, such as ambient temperature, grouping factor, diversity factor and the type of circuit protection. In a domestic installation, final circuits can be protected by miniature circuit breakers

(BS 3871), cartridge fuses (BS 1361) or rewirable fuses (BS 3036). The first protective device has a two-operational function based on thermal overload protection and magnetic short circuit protection. The second device has a superior short circuit rating against faults whilst the third device is the least expensive, but suffers a de-rating factor of 0.725 because its fusing factor exceeds 1.45.

3 Having decided to install a twin PVC insulated and sheathed cable, the following procedure is used to select an appropriate size:

a) determine the design current and select an appropriate protective device;
b) determine the effective current carrying capacity of the cable;
c) consult Appendix 4 of the *IEE Wiring Regulations* and the relevant tables;
d) choose a suitable size cable from Table 4D2A
e) check the cable's voltage drop against the stated percentage given in Reg 525–01–02;
f) check to see if the final circuit will disconnect in the time of 5 s (see Reg Table 41C and calculate the actual circuit impedance);
g) check the thermal constraints of the internal protective conductor (see Reg 543–01–01).

Note: Cable selection procedures can be found in chapter 3, on supply and distribution of electricity of the author's *Part 1 Studies: Theory* book.

2.8 Voltage and current measurement in a radial circuit

Task aims

To measure current, terminal voltage and voltage drop in a radial circuit incorporating several load positions.

Task objectives

You should:

1 study Figure 2.11 and connect the circuit as indicated;
2 take readings of the ammeters and voltmeters in different parts of the circuit;
3 determine the voltage drop at each load point.

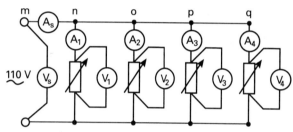

Figure 2.11 Measurement of current and voltage in a radial circuit

Observed competences

The person supervising your practical work will be looking for the following points:

1 Your ability to connect ammeters and voltmeters in a radial circuit.
2 Your ability to record accurately current and voltage in different parts of the circuit.
3 Your ability to calculate voltage drop at different load positions.

Guidance notes

1 Before switching on the circuit in Figure 2.11, check to see if the ammeters and voltmeters are properly zeroed. Most analogue instruments, especially multimeters, incorporate a mirror to eliminate reading error. If you look directly over the pointer of such an instrument and no reflection is seen on either side, an accurate reading will be taken.

2 Set up the circuit according to the following procedure, switch on the supply and take readings of current and voltage.
3 *Load setting 1*

   a) adjust each load resistor so that A1 reads 2.2 A, A2 reads 3.8 A, A3 reads 1.5 A and A4 reads 2.5 A;
   b) using the voltmeter, measure the voltage across the supply terminals and each load;
   c) record the voltmeter readings in the results table Figure 2.12;
   d) determine the current in sections labelled no, op and pq.
   e) determine the voltage drop at the load positions.

4 *Load setting 2*

   f) re-adjust each load resistor so that A1 reads 5 A, A2 reads 1 A, A3 reads 3 A and A4 reads 6 A;
   g) repeat the above procedure from step (b).

5 Assuming that the maximum voltage drop allowed is 4%, determine the percentage voltage drop at each load position.
6 Write comments about the current distribution, the terminal voltage of each load and the voltage drop in different parts of the circuit.
7 State Kirchhoff's first law of current distribution. Note: This law is explained and used in chapter 2, on basic circuit theory of the author's *Electrical Installation Technology 2* book.

Circuit condition	V_s	V_1	V_2	V_3	V_4	A_s	A_1	A_2	A_3	A_4	I_{no}	I_{op}	I_{pq}
Load setting 1	110 v					10	2.2	3.8	1.5	2.5			
Load setting 2	110 v					15	5	1	3	6			

Voltage drop (V) $V = V_s - V_1$	Percentage voltage drop $X\% = \dfrac{(V_s - V_1)100}{V_s}$

Figure 2.12 Results tables

2.9 Three-phase, cage induction motor test

Task aims

1. to connect a three-phase cage induction motor to a direct-on-line starter;
2. to control the operation of the motor using remote stop and start buttons;
3. to show how the direction of rotation of the motor can be reversed.

Task objectives

You should:

1. study Figure 2.13 and connect the starter to the motor;
2. switch on the supply and start the motor running, observing the direction of rotation;
3. reconnect the motor for remote operation using additional stop and start buttons;
4. reverse the motor's direction of rotation by modifying the circuit wiring.

Observed competences

The person supervising your practical work will be looking for the following points:

1. Your ability to study a circuit diagram and connect a three-phase, cage induction motor to a direct-on-line contactor starter.
2. Your ability to modify the motor circuit with stop and start buttons for remote operation.
3. Your ability to change the rotational direction of a three-phase motor.

Guidance notes

1. Figure 2.13 shows a complete circuit diagram of the motor and starter. For remote operation you will see that start buttons are connected in parallel while stop buttons are connected in series. Figure 2.14 shows a typical starter circuit and the modifications needed to it for remote stop and start operation.
2. You will find a description of the operation and starting methods of cage induction motors in chapter 3, electric motors and starters of the

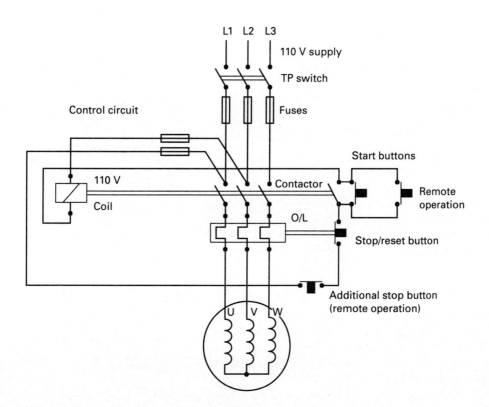

Figure 2.13 3-phase motor with direct-on-line starter circuit

author's *Part 2 Studies: Science* book. Here, you will discover why the d.o.l. starter has overload and no-volt protection. You will also find out how to reverse the motor's rotation by changing over two stator supply connections.

3 In carrying out your wiring, it is important to check connections but not to over-tighten terminal screws. You should also make sure of the earth continuity between the motor and starter, remembering that the motor will vibrate once it is started.

4 You should take the opportunity in this task to measure the motor's speed using a stroboscope and determine its percentage or per unit slip. Also, with a suitable ammeter connected in one of the lines, you should observe the in-rush current to the motor's windings when it is first switched on.

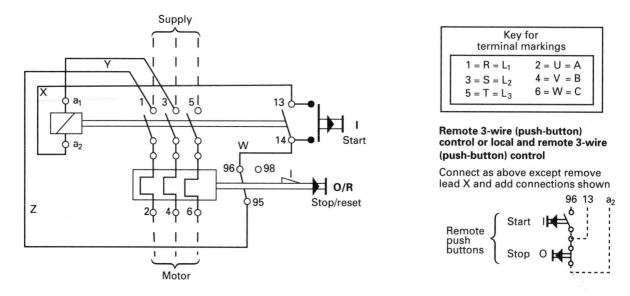

Figure 2.14 Facilities for controlling a 3-phase motor using remote push buttons (courtesy of MEM Co. Ltd.)

2.10 Dismantling and reassembling a three-phase motor

Task aims

To dismantle a three-phase cage induction motor, sketch its component parts and then reassemble it for operation on the a.c. supply.

Task objectives

You should:

1. study Figure 2.15 and identify the component parts of the motor;
2. dismantle the motor with the aid of the correct tools and reference to guidance notes;
3. sketch the main parts of the motor, namely its stator, rotor and end shields;
4. reassemble the motor and check its operation by connecting it to the a.c. supply.

Observed competences

The person supervising your practical work will be looking for the following points:

1. Your knowledge of the motor's operation and component parts.
2. Your ability to dismantle the motor using the correct tools.
3. Your ability to sketch the motor's main component parts.
4. Your ability to re-assemble the motor and check its operation for running correctly.

Guidance notes

1. If the motor is fitted with a pulley, it needs to be removed by extracting the taper gib on the shaft's keyway using a taper drift. If, however, a taper lock is fitted a hexagon wrench must be used. Hub screws are inserted in jacking-off holes and as they are tightened the bush in the hub becomes free on the shaft.

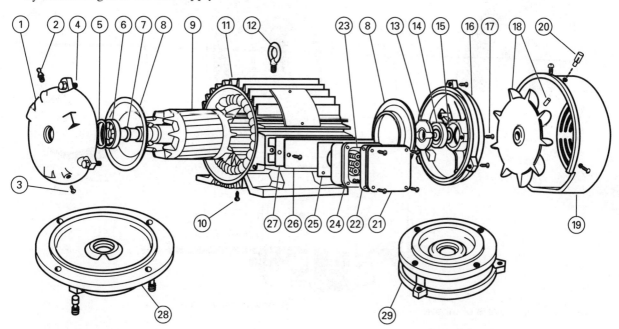

1. Endshield, driving end
2. Grease nipple
3. Grease relief screw
4. End securing bolt, or through bolt and nuts
5. Anti-bump washers
6. Ball bearing - driving end
7. False bearing shoulder
8. Flume
9. Rotor on shaft
10. Drain plug
11. Yoke with or without feet
12. Eyebolt
13. Inside cap, non-driving end
14. Ball bearing, non-driving end
15. Circlip
16. Endshield, non-driving end
17. Inside cap screws
18. Fan with peg or key
19. Fan cover
20. Lubricator extension plug
21. Terminal box cover
22. Terminal box cover gasket
23. Terminal board
24. Terminal box
25. Terminal box gasket
26. Raceway plate
27. Raceway plate gasket
28. D flange
29. C face flange

Figure 2.15 Components of a typical cage-rotor induction motor

2 With reference to Figure 2.15, unscrew the fixing bolts of the non-drive endshield (item 16) and remove it from the motor using a mallet and wooden block. Repeat this procedure for the drive endshield and extract the cage assembly. Note: If the motor has bearing grease covers on the outside of the endshields, they should be removed first.

3 Inspect the internal stator, looking for any possible damage caused by hot spots.

4 Sketch the component parts of the motor (see Figure 2.16).

5 If you want to remove the drive side bearing, you will need an extracting tool. Firstly, measuring the distance the bearing is from the end of the shaft. Apply the extracting tool and remove the bearing.

6 Remove the existing grease in the bearing cover and replace with fresh grease.

7 Slide the replacement bearing on the shaft and use a drift tool to knock the bearing down to the required distance.

8 Insert the cage back into the motor and use threaded studs to locate the drive side end shield and bearing cover.

9 Repeat the procedure for the assembly of the non-drive endshield.
Note: You must not use a steel hammer to tap the end shields.

10 Check the free running of the shaft when the assembly is complete, then test the motor's windings for continuity and insulation resistance.

11 Connect the motor to its starter and run it up to its full rated speed. Observe its performance.

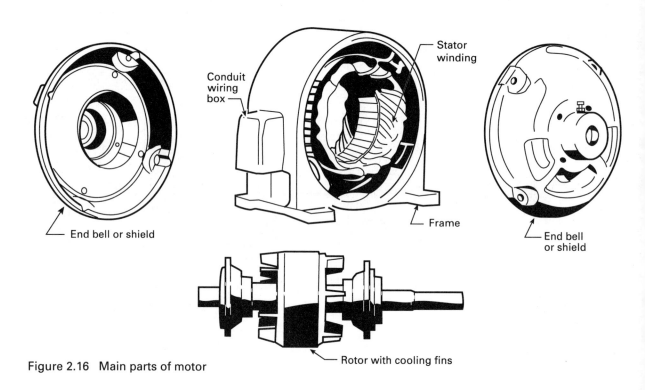

Figure 2.16 Main parts of motor

2.11 Operation of a contactor

Task aims

To install a triple-pole contactor, operating a three-phase load through normally open and closed contacts.

Task objectives

You should:

1. study the circuit diagram in Figure 2.17 and connect the circuit as shown;
2. test the circuit for its correct working;
3. describe how the circuit operates, stating the two main functions of the operating coil circuit;
4. state the requirements of the *IEE Wiring Regulations* concerning protection against undervoltage.

Observed competences

The person supervising your practical work will be looking for the following points:

1. Your ability to read a diagram of a contactor control circuit.
2. Your ability to connect the circuit for its proper working.
3. Your understanding of how the circuit operates.
4. Your understanding of the regulation requirements for protection against undervoltage.

Guidance notes

1. Study Figure 2.17 and identify the circuit components.
2. Obtain the necessary materials and proceed to install and wire the components in the positions shown.
3. When the wiring is completed, check the circuit, switch on the supply and test for correct operation.
4. Briefly describe how the circuit operates. Note that the purpose of the coil is to hold in the contactor. Once the supply is cut off, it cannot be restored to the load again until the contactor coil is energised via the start buttons.
5. The control circuit serves the same purpose as a direct-on-line starter but does not have overload protection – refer to regulation 552–01–01 of the *IEE Wiring Regulations*.
6. For the requirements concerning protection against undervoltage, see chapter 45, section 451 of the *IEE Wiring Regulations*.

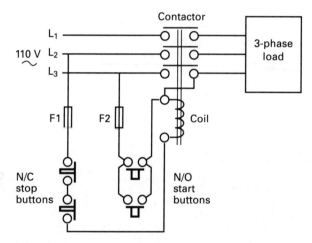

Figure 2.17 Contactor circuit

2.12 Lighting circuits

Task aims

To wire and connect the following types of lighting circuit:

incandescent tungsten lamp (GLS);
tungsten halogen lamp (HL);
low pressure mercury lamp (MCF);
high pressure mercury lamp (MBF);
low pressure sodium lamp (SOX);
high pressure sodium lamp (SON).

Task objectives

You should:

1 study the circuit diagrams in Figure 2.18 and connect the lamps and their associated control gear as shown;
2 test the lamp circuits for their correct operation;
3 briefly describe how the circuits operate, stating the main function of any control gear.

Observed competences

The person supervising your practical work will be looking for the following points:

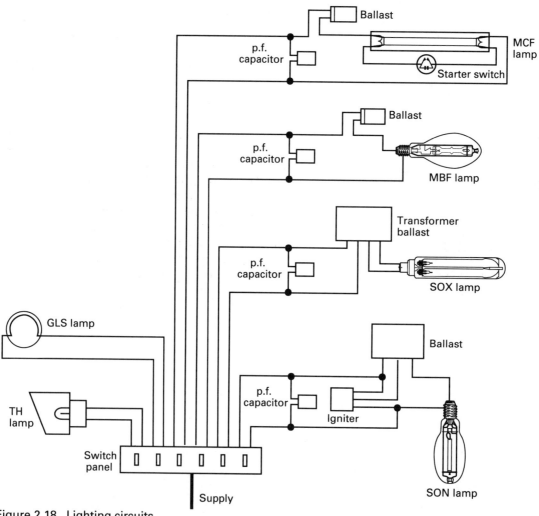

Figure 2.18 Lighting circuits

1 Your ability to read circuit diagrams of the lamps listed.
2 Your ability to connect and test for correct working of the lamps listed.
3 Your general understanding of the operation of the lamps listed.

Guidance notes

1 Study the lamp circuits shown in Figure 2.18, noting the connections of circuit components.
2 Connect each lamp circuit to the supply and test it for its correct operation. Note the colour appearance of each lamp as it reaches full brightness.
3 With regard to the MFC, MBF, SOX, and SON lamps these should be switched on and off to see if they are able to re-strike immediately.
4 You will find a description of all the lamps listed in the author's *Part 1 Studies: Theory* book, Chapter 3, pages 51–59 and also in the author's *Part 2 Studies: Science* book, Chapter 4, pages 73–77.

2.13 Current measurement in a SON discharge lamp

Task aims

To measure the circuit currents in a SON discharge lamp with and without the power factor improvement capacitor connected.

Task objectives

You should:

1. study the circuit diagram in Figure 2.19 and connect the ammeters where shown;
2. measure the currents in the circuit with the power factor capacitor switched in and out of circuit;
3. comment on the ammeter readings and reasons for including a capacitor in the circuit.

Observed competence

The person supervising your practical work will be looking for the following points:

1. The correct wiring of Figure 2.19.
2. Your ability to measure and record currents values.
3. Your understanding of why the supply current ammeter value changes.

Guidance notes

1. Study the circuit diagram of Figure 2.19 and then make reference to Chapter 4, exp. 4.6, page 74, showing an investigation into power factor improvement for a SOX lamp.
2. When you have wired the circuit and switched on the supply, you will find that the capacitor causes less current to flow through the supply ammeter. This improvement reduces the lamp circuit's running costs.
3. A typical SON discharge lamp question appears in the author's *Question and Answers* book, Q198 on page 116. You will find a description of power factor in the author's *Part 2 Studies: Science* book, pages 33–34.

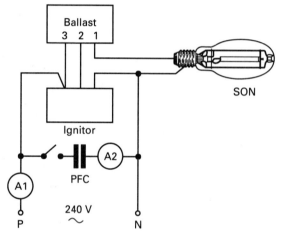

Figure 2.19 Measuring current in a SON lamp circuit

2.14 Voltage measurement in a fluorescent lamp circuit

Task aims

To measure the voltage across different parts of a fluorescent lamp circuit.

Task objectives

You should:

1. study the circuit diagram in Figure 2.20 and connect the voltmeter across the circuit where requested;
2. record your voltage readings in the results table;
3. comment on the voltages found in different parts of the circuit.

Observed competences

The person supervising your practical work will be looking for the following points:

1. Your ability to connect safely a voltmeter in different parts of the fluorescent lamp circuit.
2. Your ability to record and tabulate voltmeter readings.
3. Your understanding of circuit components.

Guidance notes

1. Study Figure 2.20 and note the positions for making the voltage tests requested in Figure 2.21.
2. Obtain a proprietary voltage tester (digital or analogue) and on a suitable voltage range, test across parts of the fluorescent lamp circuit where indicated.
3. Insert your voltage readings in Figure 2.21 and briefly describe how the circuit components operates. **Note:** You should make reference to the author's *Part 1 Studies: Theory* book, pages 53–55 for an explanation of the lamp's control gear.

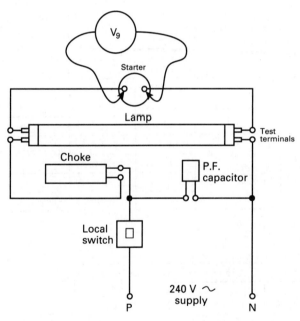

Figure 2.20 Voltage measurement in a fluorescent lamp circuit

Circuit components to test		Voltage
Supply on choke or ballast to neutral	V_1	
Output on choke or ballast to neutral	V_2	
Starter side of tube end No. 1 to neutral	V_3	
Starter side of tube end No. 2 to neutral	V_4	
Starting voltage across tube (running)	V_5	
Choke input to output	V_6	
Filament voltage tube end No. 1	V_7	
Filament voltage tube end No. 2	V_8	
Across glow starter switch terminals (running)	V_9	

Figure 2.21 Voltage test results table

2.15 Test instruments

Task aims

To carry out tests and take readings using the following measuring instruments:

milliohmmeter;
multirange instrument;
insulation/continuity tester;
earth fault loop impedance tester;
residual current device tester.

Task objectives

You should:

1 study the diagram in Figure 2.22 and take measurement of the electrical quantities listed;
2 exercise care when using the listed instruments and record your results in the table provided (see Figure 2.23);
3 confirm your test results with the model answer sheet.

Observed competences

The person supervising your practical work will be looking for the following points:

1 Your ability to use correctly and safely the instruments listed.
2 Your ability to record and tabulate instrument readings.

Guidance notes

Study the diagrams shown in Figure 2.22 and then make the following tests. Record your results using a table as in Figure 2.23.

1 Using the milliohmmeter, determine the resistance values of the three resistors in Exercise 1.

	Exercise and instrument used	Quantity being measured	Point of measurement	Test results
1	Milliohmmeter	Resistance	A	____ Ω
		Resistance	B	____ Ω
		Resistance	C	____ Ω
2	Multimeter	Resistance	A	____ Ω
		Resistance	B	____ Ω
		Resistance	C	____ Ω
		Voltage	V	____ V
		Current	A	____ A
3	Continuity and insulation resistance tester	Resistance	P	____ Ω
		Resistance	N	____ Ω
		Resistance	cpc	____ Ω
		Resistance	P – N	____ MΩ
		Resistance	PN – cpc	____ MΩ
4	Loop impedance tester	Impedance	A	____ Ω
		Impedance	B	____ Ω
		Impedance	C	____ Ω
5	Residual current device tester	Time	at 15 mA	____ s
		Time	at 30 mA	____ s
		Time	at 150 mA	____ s
		Time	Test button	Fast/slow

Figure 2.23 Test results table

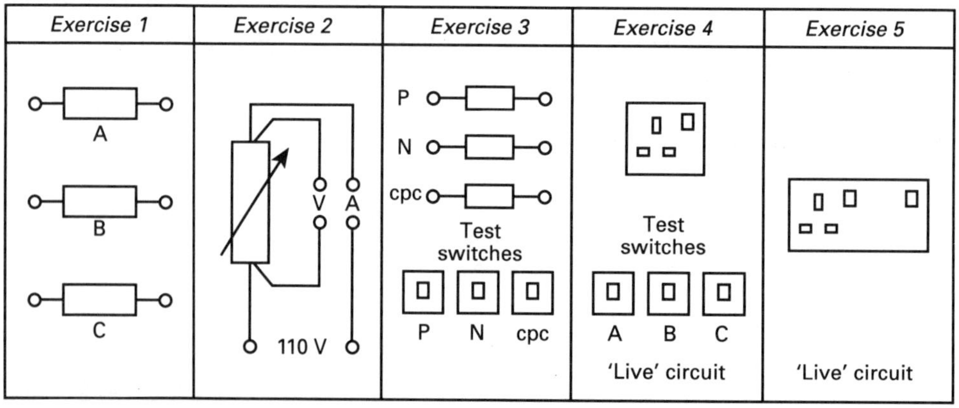

Figure 2.22 Test simulator board

2 Repeat the test in Exercise 1 using the multimeter set on the correct resistance range (see the instrument's working instructions). Continue to use the multimeter to find the current and voltage in Exercise 2, then record your results.

3 Use the insulation/continuity tester (ohmmeter) to find the resistance values in Exercise 3. Set the test switches and then record your results.

4 Use the earth fault loop impedance tester to find the ohmic impedance of the socket outlet in Exercise 4. Take readings for each test switch and then record your results.

5 Use the residual current device test instrument to test the r.c.d. in Exercise 5 and then record your results.

2.16 Testing an electrical installation

Task aims

To carry out prescribed tests on an electrical installation in accordance with the *IEE Wiring Regulations*.
Note: this does not include testing for ring circuit continuity which is conducted later under fault finding.

Task objectives

You should:

1. study the diagram in Figure 2.24 and with the correct instrument make the following tests:
   a) continuity of protective conductors;
   b) insulation resistance;
   c) polarity;
   d) earth electrodes resistance;
   e) earth fault loop impedance;
   f) residual current device;
2. record your results in the table provided (see Figure 2.26).
3. comment on your results, stating whether or not the requirements of the IEE Wiring Regulations have been satisfied.

Observed competences

The person supervising your practical work will be looking for the following points:

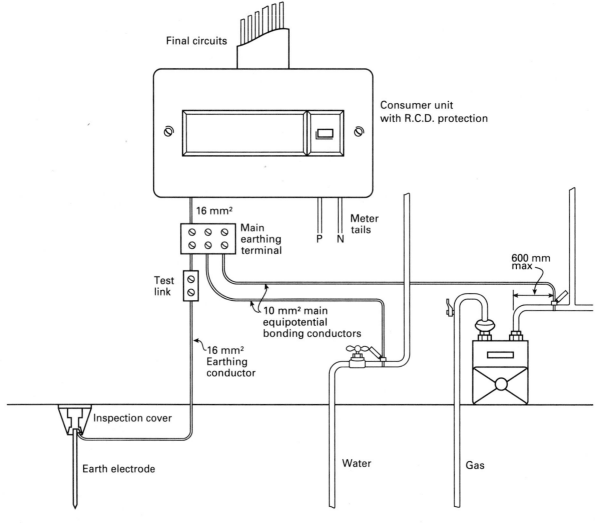

Figure 2.24 Consumer's intake position

1. Your ability to use safely and correctly the test instruments.
2. Your ability to record and tabulate the test results.
3. Your understanding of the *IEE Wiring Regulations* regarding the testing of the electrical installation.

Guidance notes

1. Study Figures 2.24 and 2.25 and then carry out the following test on the final circuits:

   a) test the continuity of all the circuit protective conductors using the ohmmeter on its low resistance range and record your results;
   b) test the complete wiring for insulation resistance between all live conductors and between all live conductors and earth using the ohmmeter on its high resistance range;
   c) determine the polarity of the phase conductors at switch positions and lighting points and also determine the polarity of connections at every socket outlet, using the ohmmeter on its low range for this test;
   d) test the earth electrodes using a proprietary earth electrode resistance tester following the instructions on how to use the tester;
   e) with the supply available, make an earth fault loop impedance test at the furthest point of each final circuit and also, make a test on the residual current device using a proprietary r.c.d. tester.

2. To successfully carry out the above tests, you should make reference to the *IEE Wiring Regulations, Guidance Notes No.3 'Inspection and Testing'*, section 17, page 89. You should also refer to the author's *'Part 1 Studies: Theory'* book, chapter 4 inspection and testing pages 77–95. Record your results in a table similar to Figure 2.26.

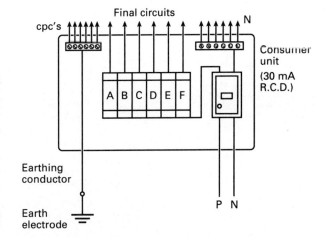

Figure 2.25 Details of final circuit arrangements to Figure 2.24

Circuit protective devices		
A	Lighting	5 A
B	Lighting	5 A
C	Heater	15 A
D	Ring	30 A
E	Ring	30 A
F	Cooker	45 A

Test methods	Test results	
Continuity of protective conductors	Lighting circuit	
	Lighting circuit	
	Heater circuit	
	Ring circuit	
	Ring circuit	
	Cooker circuit	
Insulation resistance	Between live conductors	
	Between live conductor and earth	
Polarity	Switches	
	Ceiling rose	
	Socket outlets	
Earth electrode resistance	Earth electrode	
Earth fault loop impedance	Lighting circuit	
	Lighting circuit	
	Heater circuit	
	Ring circuit	
	Ring circuit	
	Cooker circuit	
Residual current device	50% trip current	
	100% trip current	
	150 mA test	
	Test button	

Figure 2.26 Test results table

2.17 Testing electronic equipment

Task aims

To carry out tests and take readings on electronic equipment using the following instruments:

a) digital and analogue multimeters;
b) signal generator;
c) cathode ray oscilloscope;
d) logic probe;
e) frequency meter.

Task objectives

You should:

1 with reference to Figure 2.27 study the user instructions of each instrument listed above and carry out tests on selected electronic components;
2 record test readings and comment on your results.

Observed competences

The person supervising your practical work will be looking for the following points:

1 Your understanding of how to use the listed instruments.
2 Your ability to safely and correctly use the instruments on circuit components and record test results.

Guidance notes

1 Study Figure 2.27 and make tests with the selected instruments. You should be extra careful using the instruments, thoroughly checking the connections and making sure of the correct range settings.
2 Record your test results and write brief notes on your findings.
3 See page 91 for further information.

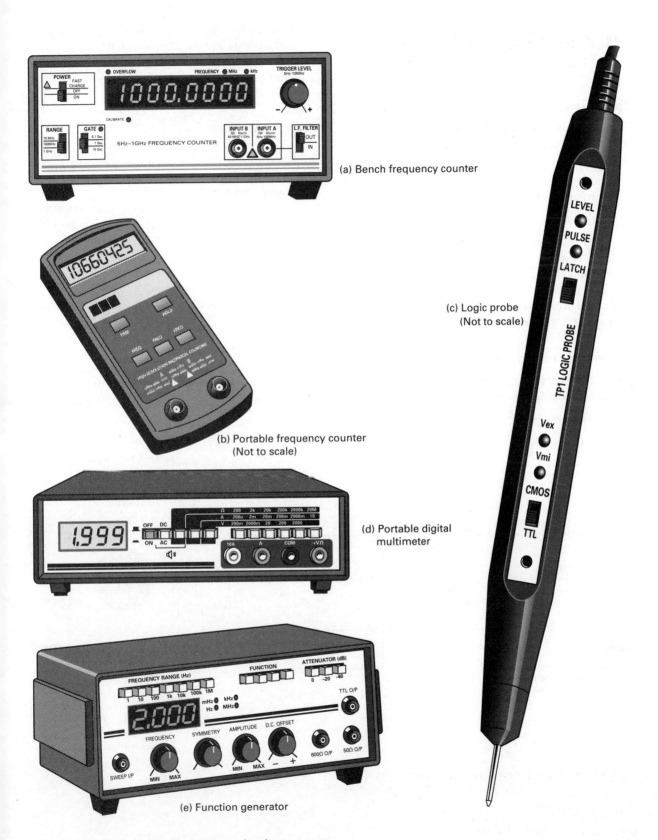

Figure 2.27 Typical electronic measuring instruments

2.18 Measurement of amplitude and frequency

Task aims

To observe the input and output waveforms of a simple audio amplifier using a frequency generator, voltmeter and cathode ray oscilloscope.

Task objectives

You should:

1. study the circuit diagram shown in Figure 2.28 and connect the instruments in circuit to measure the amplitude and frequency of both input and output waveforms;
2. record your results in the table and make comments on your findings.

Observed competences

The person supervising your practical work will be looking for the following points:

1. Your ability to use correctly a signal generator and cathode ray oscilloscope (CRO).
2. Your ability to measure frequency and waveform on an oscilloscope.

Guidance notes

1. Read the instruction notes on using the CRO found in Chapter 5 and then connect the instruments shown in Figure 2.28.
2. Connect the signal generator and voltmeter across the input of the amplifier. Apply 10 V d.c. to the circuit and set the signal generator's frequency to 1 kHz at 30 mV. Set up the oscilloscope (as explained in Chapter 1) and measure the frequency and amplitude.
3. Reconnect the oscilloscope to the output of the amplifier and observe if the signal is sinusoidal.
4. Raise the input to 50 mV and note the effect.
5. Sketch the input and output waveform.
6. Measure the amplitude and frequency as explained in Chapter 5 and record the results.

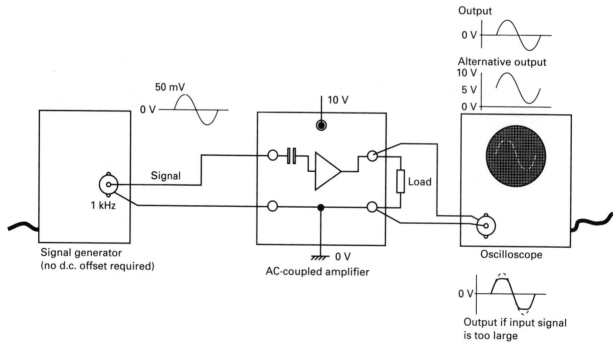

Figure 2.28 Waveforms of a simple audio amplifier

2.19 Fault finding in a ring circuit

Task aims

To diagnose numerous faults found in the wiring of a 30 A ring final circuit shown in Figure 2.29.

Task obectives

You should:

1. Study the diagram shown in Figure 2.29 and with the correct instrument make the following tests:

   a) continuity of the live and protective conductors of the circuit;
   b) insulation resistance between live conductors and between live conductors and earth;
   c) polarity at each socket outlet.

2. Record your results in the table provided (see Figure 2.30).
3. Comment on your results, stating the requirements of the IEE Wiring Regulation.

Test description of ring circuit	Fault switches	Diagnosis
Continuity resistance phase conductor	1	
Neutral conductor	2	
Circuit protective conductor	3	
Insulation resistance between phase and neutral	4	
Between live conductors and cpc	5	
Polarity	6	
S.O. 1 – L. N. E. S.O. 2 – L. N. E.		
S.O. 3 – L. N. E. S.O. 4 – L. N. E.		
S.O. 5 – L. N. E. S.O. 6 – L. N. E.		
S.O. 7 – L. N. E. S.O. 8 – L. N. E.		
Comments		

Figure 2.30 Results table

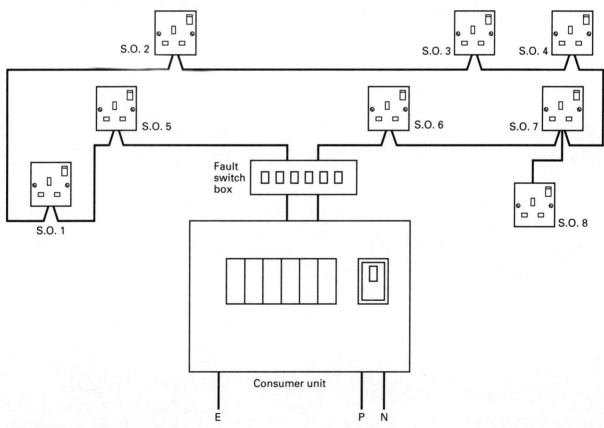

Figure 2.29 30 A ring final circuit

Observed competences

The person supervising your practical work will be looking for the following points:

1 Your ability to use safely the test instruments.
2 Your ability to record and tabulate test results.
3 Your understanding of the faults and their possible cause.
4 Your understanding of the requirements of the *IEE Wiring Regulations* regarding the tests made.

Guidance notes

1 Study Figure 2.29 and make tests on the ring final circuit in the sequence listed above.
2 Test the circuit's earth continuity resistance, insulation resistance and polarity in accordance with the *IEE Wiring Regulations*, Guidance Notes No. 3.
3 Record your results in the table (see Figure 2.30) and make comments on the faults found, stating their possible cause and remedy for compliance with the *IEE Wiring Regulations*.

2.20 Fault finding in a lighting circuit

Task aims

To diagnose numerous faults found in the wiring of a high-pressure sodium lamp circuit.

Task objectives

You should:

1. study the circuit shown in Figure 2.31 and verify that faults exist;
2. list the possible causes, remedy the fault and test the circuit;
3. operate the lamp over its run-up time.

Observed competences

The person supervising your practical work will be looking for the following points:

1. Your ability to diagnose the lamp circuit faults.
2. Your ability to test and commission the lamp circuit once the faults have been found.

Guidance notes

1. Study Figure 2.31, switch on the supply and make observations. Look for the following symptoms:

   a) lamp is intact but does not light;
   b) low light output;
   c) flickering light;
   d) lamp keeps cycling on and off.

2. In (a), try another lamp and if this is not the problem, check the supply voltage and the connections of circuit components;
   in (b), check the supply for the correct voltage;
   in (c), check all the circuit connections;
   in (d), replace the lamp.

3. Once the faults have been corrected, operate the lamp for 5 minutes.

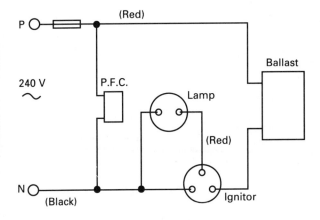

Figure 2.31 Circuit diagram of a SON discharge lamp

Course assignments 3

Objectives

After working through this chapter you should be able to:

- *interpret designer's drawings which use BS 3939 graphical symbols;*
- *draw different types of circuit wiring diagram;*
- *construct layout, block and 'as fitted' diagrams;*
- *produce material requisitions from drawings;*
- *select conduits and trunking sizes from tables;*
- *design a simple lighting scheme;*
- *state numerous requirements of the IEE Wiring Regulations.*

3.1 Introduction

As part of the electrical installation competences course, the examining body, City and Guilds of London Institute, set assignments for Part I and Part II students to complete. These assignments are compulsory and are graded a pass or fail by the students' college tutor.

Part I students have to complete three assignments, each of approximately 5 hours duration and Part II students have to complete one assignment of approximately 30 hours duration.

The Part II assignment is component 236–8–21 of Unit 1 *Communication and Industrial Studies*. It covers the following six sections:

A – structure of the industry
B – contract administration
C – health and safety
D – site administration
E – installation drawings
F – electronics

In *Section A*, you are expected to state the purpose of national and local agreements, contracts of employment, grading schemes, functions of employers' associations, trade unions and the Joint Industry Board. You also have to recognise the procedures for settlement of disputes. A typical assignment question in this area might be: 'An electrician employed by your company has not been given a contract of employment. State how this might affect him.'

In *Section B*, you are expected to state the relationship between the main building contractor and the electrical subcontractor. In this section you might be asked to: 'List several ways in which you would try to establish and maintain a good customer relationship with the main contractor.'

In *Section C*, you are expected to identify a number of statutory and non-statutory regulations, state the scope of the *Health and Safety at Work Act* and prepare a mock accident report. Here you could be asked to: 'Explain the correct action to be taken at the scene of an accident in which an electrician has fallen from a pair of steps and is suffering concussion and a suspected broken arm.

In *Section D*, you are expected to interpret bar charts to determine the sequence of work for given situations. You have to state the need for formal contract procedures and explain the need for good industrial relations and good customer relations. You also have to identify and complete specimen copies of day work sheets, job sheets, time sheets and delivery records, etc. A typical question in this section could be: 'The main contractor has issued a variation order for you to install an extra socket outlet in each room. Prepare a day work sheet for this extra work.'

In *Section E*, you are expected to read and interpret designers' drawings which use BS 3939 circuit symbols. You also have to prepare block and circuit diagrams as well as 'as fitted' drawings. From the given drawings of an installation, you have to prepare a material requisition, and by interpretation of the specification you will have to identify suitable tools and equipment needed to carry out the wiring. A question in this section might be: 'On an overlay, prepare an "as fitted" drawing for the lighting installation.'

In *Section F*, attention is given to your recognition of BS 3939 symbols for electronic components, such as diodes, transistors, etc. You are expected to state different types of electronic drawing and state the reasons for the use of component positional reference systems. Here you might be asked to: 'Draw the circuit diagram shown in Fig 000–00 on to a grid to produce a component positional diagram.'

In this chapter we will look at a number of practical examples which concentrate around Section E. Information about the other sections can be found in the author's *Part 2 Studies: Theory* book in this series and the author's *Electrical Installation Technology 1: Theory and Regulations* book. The first two examples are from the latter book and have been modified.

It is the intention that you (the reader) participate in the examples where indicated under the subheading of student activity. At the end of each example, the person supervising your classwork will be looking for one or more of the following points:

- production of a traced overlay;
- list of materials used (types of accessory to be indicated);
- diagrams, fully labelled and neatly drawn, showing where possible how the circuit operates;
- comments and regulation requirements where indicated.

3.2 Section E – installation drawings

Example 1

Figure 3.1 is a room in part of a light *engineering workshop* which is supplied with its own three-phase distribution board.

a) Identify all the BS 3939 installation location symbols.
b) On an overlay, prepare an 'as fitted' drawing of the lighting installation.
c) From the drawing, produce a material requisition in order to carry out the erection of the conduit and the wiring.
d) Draw a wiring diagram of the lighting circuit assuming three final circuits from the same phase.
e) If there are two other distribution boards in the engineering workshop and all three are controlled by 100 A TPN fuse switches, draw a line diagram showing the switch-gear arrangement at the intake position.

Solution

a) Student activity.
   Note: See Appendix 4, page 128 of the author's *Part 1 Studies: Theory* book.
b) See Figure 3.2
c) Student activity.
   Notes:

   1) Assume the two switch drop lengths to be 3 m.
   2) The scale 1:50 means that the drawing is one-fiftieth of its actual size.
   3) Don't list individual hand tools (say, tool box) but remember to include access equipment such as ladders and steps. Indicate the materials used with product catalogue numbers.

d) See Figure 3.3
e) See Figure 3.4

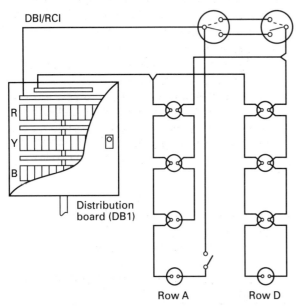

Figure 3.3 Lighting circuit for row 'A' and row 'D' (cpc's have been omitted for clarity)

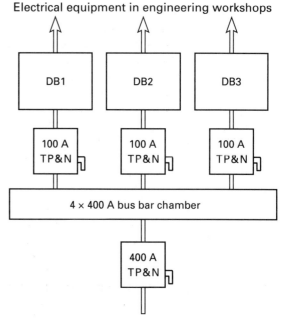

Figure 3.4 Main switchgear at intake position

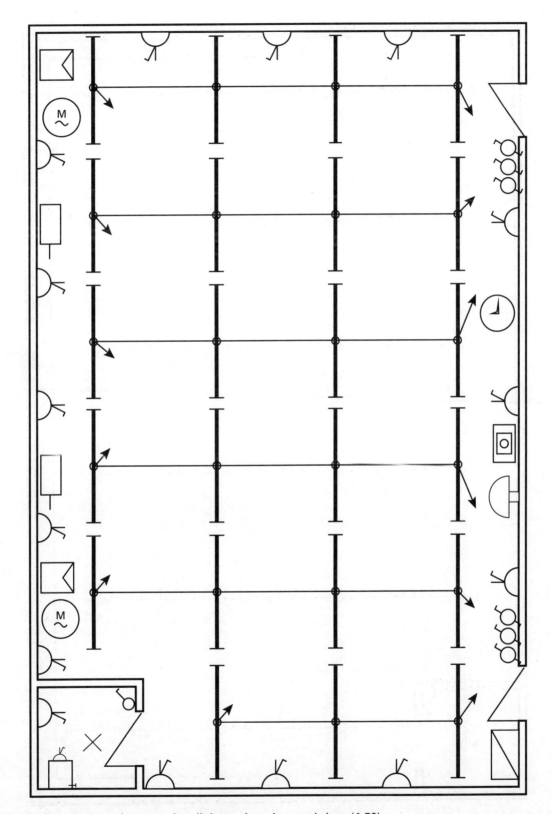

Figure 3.1 Electrical requirements in a light engineering workshop (1:50)

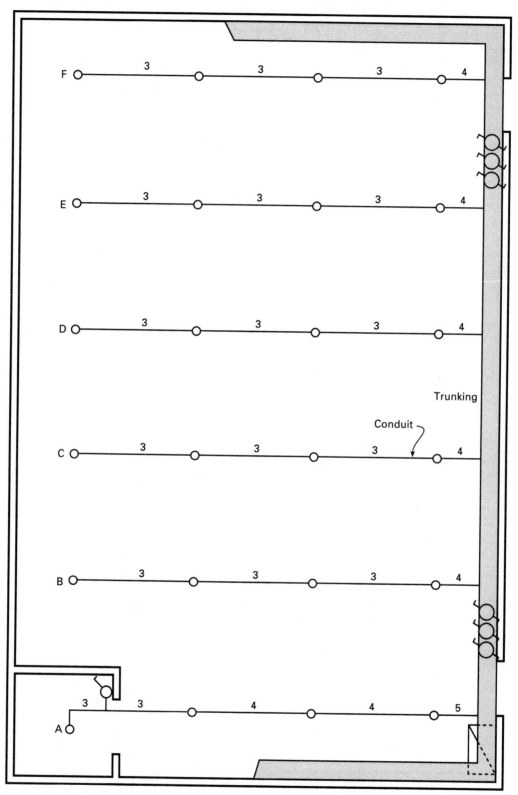

Figure 3.2 'As fitted' drawing of lighting system indicating numbers of 1.0 mm² cable in sections of conduit (1:50)

Example 2

Figure 3.5 shows the layout diagram of a *ground floor, two bedroom flat*. The floors and ceiling are of concrete construction and the walls are brickwork with a plaster finish. If the wiring system is single-core, PVC cables in 20 mm plastic conduit:

a) Produce an overlay showing the conduit wiring to the lighting, socket outlets and cooker control point.
b) Determine the approximate amount of conduit required.

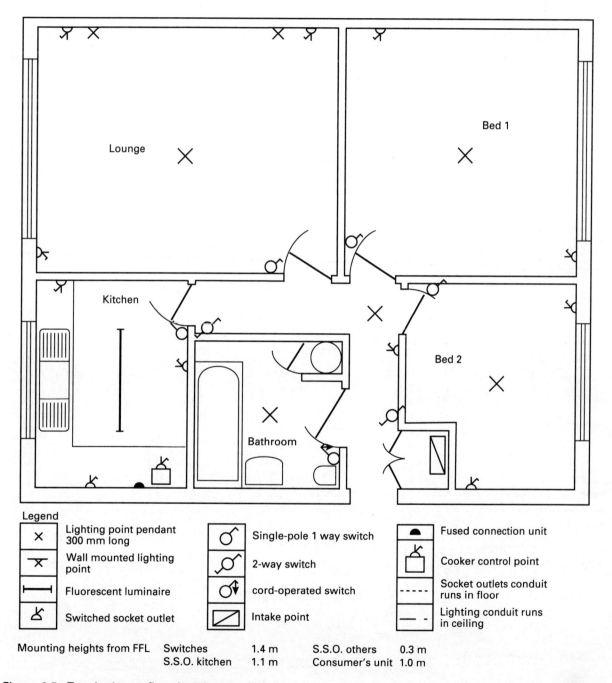

Figure 3.5 Two-bedroom flat – location symbol diagram

c) Assuming a TNC-S earthing system and a single-phase 240 V supply, draw a circuit diagram of the sequence of control at the intake position.

Solution
a) See Figure 3.6
b) Student activity.
   Note: Use a scale of 1:30.
c) Student activity.
   Note: See chapter 3, page 41 of the author's *Part 1 Studies: Theory* book.

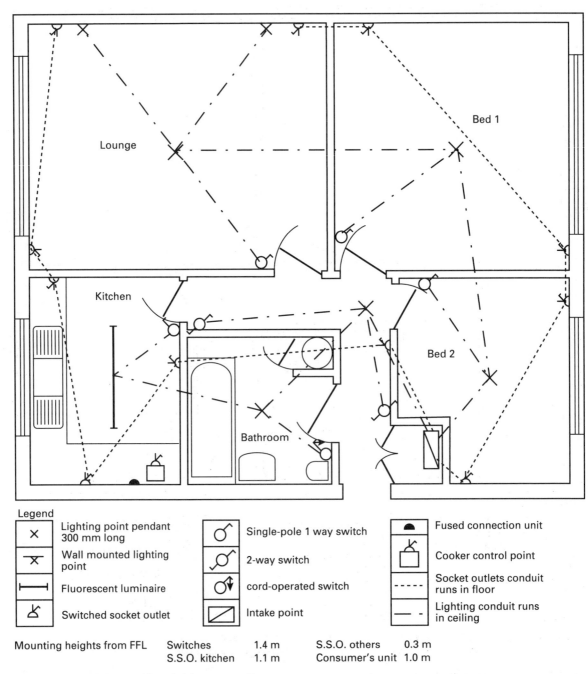

Figure 3.6 Two-bedroom flat – wiring route diagram

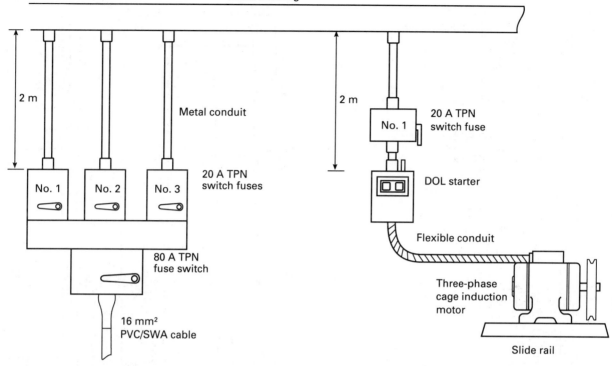

Figure 3.7 Motor installation

 Example 3

Figure 3.7 shows a sketch of a *motor installation*.

a) Make a material requisition for the installation assuming that No. 2 and No. 3 motors use the same equipment as motor No. 1. Ignore in your list the steel trunking and supply armoured cable. Assume each motor has a maximum full-load current of 8 A and the switchfuses are each fitted with 20 A Type gG, BS 88 fuses.
b) Draw a circuit wiring diagram of one of the motor's and its control gear, including the main switch.
c) State the function of the fuse switch, switch fuses and starter in respect of isolation, switching and overload protection.

 Solution

a) Student activity.
   Notes:
   1) Consult product catalogues.
   2) Complete the material list as indicated below.

Qty.	Description	Cat No.	Cost

b) See Figure 3.8.

c) See chapter 7, page 143 of the author's *Electrical Installation Technology 1: Theory and Regulations* book and also the *IEE Wiring Regulations Guidance Note No. 2 'Isolation and Switching'*.

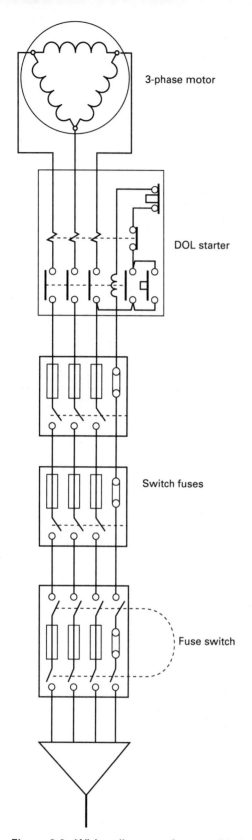

Figure 3.8 Wiring diagram of motor circuit

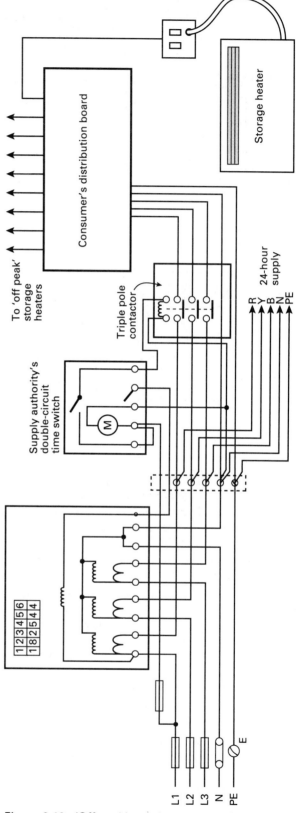

Figure 3.10 'Off peak' supply to storage heaters

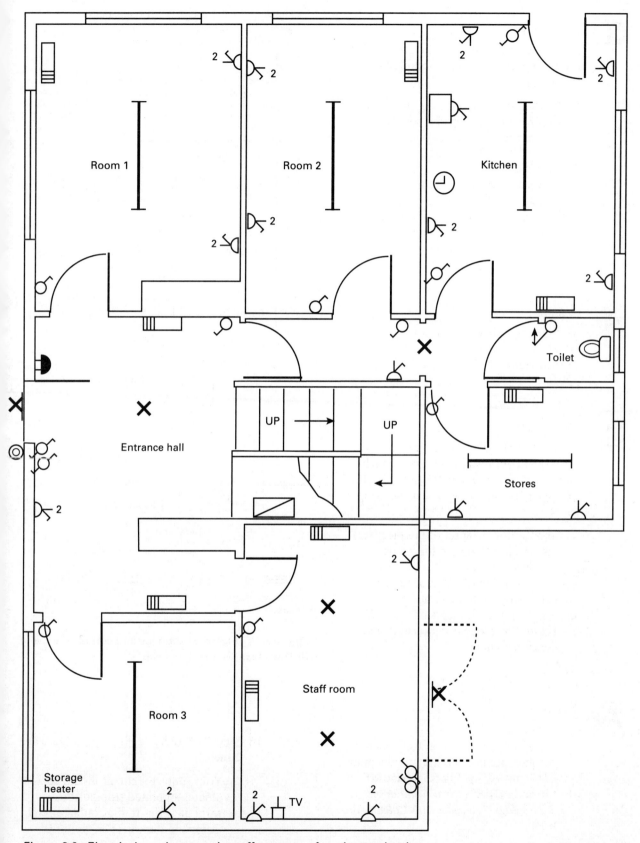

Figure 3.9 Electrical requirements in staff quarters of a private school

Example 4

Figure 3.9 shows some of the ground floor rooms of a *private school* occupied by teaching staff.

a) Prepare overlays showing the cable runs for the lighting and socket outlets – assuming the wiring system is PVC/PVC/CPC sheathed cables to BS 6004.
b) Compile a list of all the BS 3939 symbols used on the drawing.
c) Draw a circuit diagram of the storage heaters operating via a contactor, time clock and off-peak Economy 7 tariff.

Solution

a) Student activity.
b) Student activity.
c) See Figure 3.10.

Example 5

Figure 3.11 shows a *college laboratory workshop* comprising numerous bench supplies at 110 V and 240 V and also the arrangements for siting emergency stop buttons.

a) Show a line diagram of the electrical distribution to obtain the above supplies.
b) Show a circuit diagram of the emergency stop button circuit.
c) State the requirements of the *IEE Wiring Regulations* concerning emergency switching.

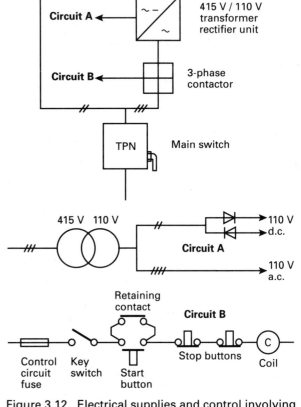

Figure 3.12 Electrical supplies and control involving RCD-protected consumer unit

Solution

a) see Figure 3.12.
   Note: To develop the transformer/rectifier circuit from the line diagram in Circuit A, see Fig 5.31, page 105 of the author's *Electrical Installation Technology 2: Science and Calculations* book.

b) Student activity.
   Note:

   1) Your contactor circuit should consist of latch-operated stop buttons wired in series to the contactor coil. Complete the wiring in circuit B.

52

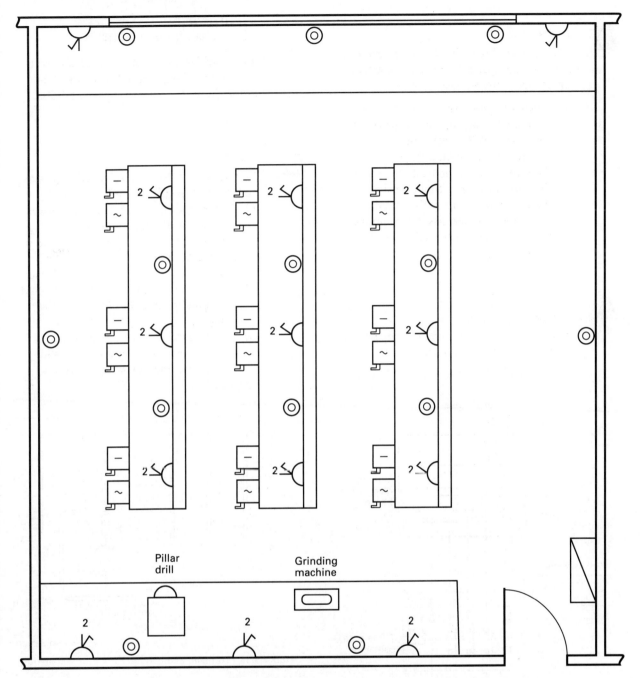

Figure 3.11 Arrangements for siting emergency stop buttons in a college electrical laboratory

2) Explain the function of the key switch and the latch-operated stop buttons.

c) Student activity.
Note: See *IEE Wiring Regulations Guidance Note No. 2*, pages 22–23.

Example 6

The lighting circuits in the *restaurant of a hotel* are controlled by two-way switching operating a three-phase contactor.

a) Draw a diagram of the circuit.
b) If the restaurant lighting consists of sixty, twin 58 W (1500 mm) fluorescent luminaires with diffusers, determine the number of circuits in the lighting distribution board and the type and rating of the circuit protective devices, assuming miniature circuit breakers are used.

 Solution

a) See Figure 3.13
b) Assume twenty luminaires per phase. Each luminaire takes a current of:

$I = (P \times 1.8)/V$
$= (2 \times 58 \times 1.8)/240 = 0.87$ A.

Five luminaires per circuit take a current of 4.35 A. This gives four circuits/phase using either 5 A or 6 A, Type 2 MCBs.

 Example 7

In Example 3, No. 2 and No. 3 motors are in the positions shown in Figure 3.14.

a) Determine the *conduit size and trunking size* for the installation, assuming that all the motor circuits are wired in stranded 2.5 mm² single-core, PVC copper cable and the trunking already carries ten 4.0 mm², eight 6.0 mm² and four mm² single-core, PVC copper cables.
b) Briefly describe with the aid of a sketch, how No. 1 motor on its slide rail can be lined up with a driven machine.

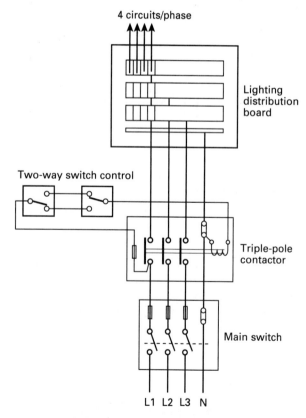

Figure 3.13 Control of lighting circuit using a contactor

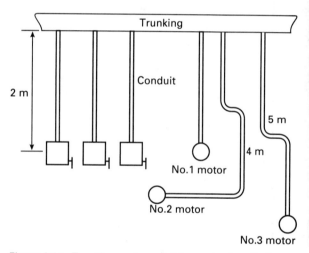

Figure 3.14 Trunking and conduit for motor installation

 Solution

a) Student activity.
 Note: Refer to Appendix 5, page 83 of the *IEE On-Site Guide*. From Figure 3.14 the 2 m lengths of conduit each containing three cables (no CPCs, and from Table 5A this gives a cable term of 129, making 16 mm conduit and 20 mm conduit suitable. From Table 5B

these conduits have capacity terms of 290 and 460 respectively. For motor No. 2 the conduit incorporates one 90° bend and therefore the conduit is regarded as having two bends. You will find in Table 5D that a 16 mm conduit has a capacity term of 130 and 20 mm conduit has a capacity term of 213 making both acceptable, although in practice 20 mm conduit would be the first choice. For motor No. 3 the conduit incorporates three bends and in this case 20 mm conduit is the minimum size acceptable. It should be pointed out that circuit protective conductors are often run in conduit systems in order to provide the system with a more reliable earth and to make impedance testing of the final circuits more individual. Having said this, you should see group regulation 522–6 of the *IEE Wiring Regulations* concerning 'impact (AG)'.

In terms of the trunking size, nine 2.5 mm² will be added to the cables already inside the trunking. The following calculations are made:

> Nine 2.5 mm² have a cable term of 102.6;
> Ten 4.0 mm have a cable term of 152;
> Eight 6.0 mm² have a cable term of 183.2;
> Four 10 mm² have a cable term of 145.2.

This gives a total cable term of 583. From Table 5F a 50 mm × 37.5 mm trunking is suitable.

b) Student activity.
   Note: See chapter 5, pages 106–107 of the author's *Electrical Installation Technology 1: Theory and Regulations* book.

Example 8

a) With the aid of a circuit diagram, describe a *TN-C-S earthing system*.
b) With reference to Example 2, complete the installation schedule in Figure 3.15 with relevant figures, assuming the inspection and testing results are satisfactory.
c) What is the supply authority's maximum external earth-loop impedance for a TN-C-S system?

Solution

a) Student activity.
b) Student activity.
c) Student activity.
   Note:
   1) For reference, see pages 1, 14 and 102 of your *IEE On-Site Guide*.
   2) See also Chapter 4 inspection and testing of the author's *Part 1 Studies: Theory* book.

Type of supply _____	Property _____	Contractor _____	Instruments:
Z_s at origin _____	_____	_____	RCD tester
PSSC _____ kA	_____	Test date _____	c/loop tester
		Inspection by _____	continuity tester
		Signature _____	insulation tester
			others _____

Description of work completed

Distribution board	no. of points	Fuse cb type	Fuse Rating (A)	Cable size (mm²)	Cable length (m)	CPC (mm²)	Test results					Remarks
							Z_s Ω	Ins Res MΩ	Polarity	RCD mS	Ring cont	

(Note Z_s shown as measured at outlet point.)

Main bonding check: Gas _____ Water _____

Deviations from Wiring Regulations and special notes:

Figure 3.15 Electrical installation schedule

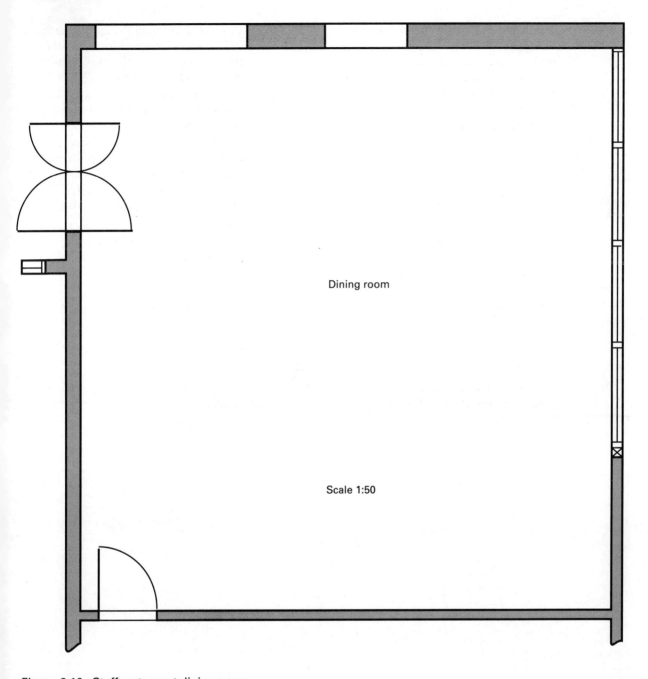

Figure 3.16 Staff restaurant dining room

 Example 9

Figure 3.16 is a *staff restaurant dining room* and is to be illuminated to a lighting design level of 200 lux. The following information is available:

    working plane 0.85 m;
    ceiling height 3.5 m;
    assumed reflectances: walls 50%, ceiling 70% and floor 20%;
    maintenance factor (light loss factor) 0.8;
    space height ratio (max), see lighting catalogue;
    metal conduit is used as the wiring system.

a) Design a suitable lighting scheme for the room.
b) Produce an overlay to show the conduit routes and show the switching arrangements, assuming two-way switching.
Note: this question is for advanced students.

Solution
a) Student activity.
   Note: See lighting manufacturers' catalogues and also chapter 4 *The elements of lighting design* in the author's *Part 2 Studies: Science* book.
b) Student activity.

Example 10
Figures 3.17 and 3.18 show the ground floor and first floor design of a *luxury 4-bedroom house*.

a) Insert in each room the appropriate BS 3939 installation graphical symbol for the lighting and small power circuits.
b) Assuming the building is of wood/brick/plaster construction and the wood joists run parallel to the side walls of the house, plan a wiring scheme in PVC/PVC/CPC sheathed cable showing the final circuit routes taken from the intake position.
c) Briefly describe some of the *IEE Wiring Regulations* covering the following:

   requirements in bathrooms;
   connections of cables at outlet points;
   earthing and bonding;
   inspection and testing.

Solution
a) Student activity.
b) Student activity.
c) Student activity.

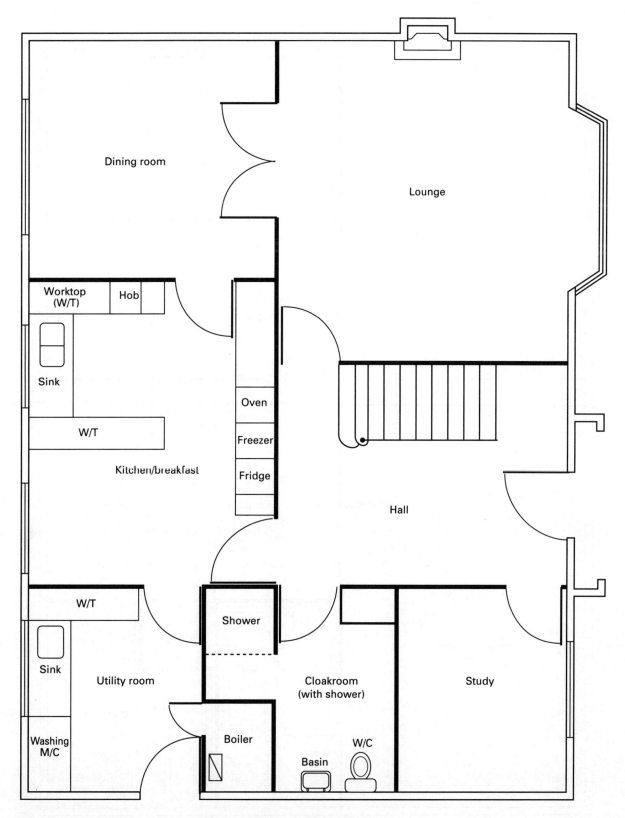

Figure 3.17 Plan view of a 4-bedroom luxury house (ground floor)

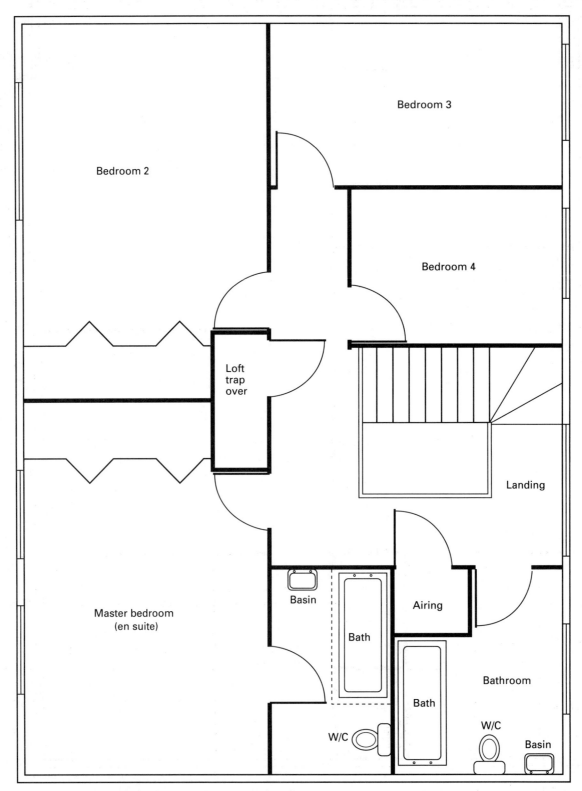

Figure 3.18　Plan view of a 4-bedroom luxury house (first floor)

Laboratory work 4

Objectives

After working through this chapter you should be able to investigate:

- *voltage and current quantities in three-phase star and delta circuits;*
- *voltage and current quantities in a.c. series circuits containing resistance, inductance and capacitance;*
- *voltage and current quantities in a series-parallel RLC circuit;*
- *characteristics and power factor of a low pressure sodium vapour discharge lamp;*
- *application of shunts and multipliers used for extending the scale range of a universal instrument;*
- *effects of variable frequencies on RLC components;*
- *speed-torque characteristics of a universal motor;*
- *methods of finding voltage drop in a cable;*
- *transformation ratios of a double-wound transformer;*
- *temperature coefficient of resistance for tungsten and carbon lamps;*
- *characteristics of a three-phase, cage rotor induction motor;*
- *power factor correction in a fluorescent lamp circuit;*
- *waveforms of half-wave and full-wave rectifiers;*
- *output waveforms of an audio frequency amplifier;*
- *use of a multimeter for measuring resistance, current and voltage;*
- *earth fault loop impedance in different parts of a circuit;*
- *starting and running performance of a 6-terminal induction motor;*
- *losses in a 1 kVA double-wound transformer.*

Guidance notes for laboratory work

This topic was introduced to you in *Practice 1* book by the same author. It outlined several important points concerning students' conduct in the laboratory and the dangers present where electricity is used; it also gave guidance on the correct procedure for carrying out experimental work.

For safety reasons, most laboratory bench supplies are provided with both 110 V a.c. and 110 V d.c. The a.c. supply is often three-phase, four-wire and 110 V is measured between lines. Between any line conductor and neutral it is 64 V. The d.c. supply is two-wire and 110 V is measured between the positive and negative poles. Both poles are fused.

When you are given an experiment to investigate you should position the apparatus in much the same way as shown on the instruction sheet. If the experiment requires a supply of electricity, you should connect the apparatus as close as possible to the supply terminals and make sure it is connected to the right type of supply, otherwise damage to the apparatus might occur. For example, a transformer connected to a d.c. supply will not produce an alternating magnetic flux to create a secondary voltage; its primary winding acts like a solenoid and by not possessing impedance the current in this winding could be excessive, sufficient to cause damage. If you are using a ring-type autotransformer, known as a variac or variable ratio transformer, you should make sure that its control knob is set at zero before switching on.

If an experiment involves a motor, which happens to be connected to some form of load, unless it is a d.c. series motor, it is important to reduce the load or remove it before switching on. This will avoid excessive inrush current to its windings which would cause them to overheat. It could also damage measuring instruments connected in circuit. Be extra careful when making connections to wattmeters, especially in three-phase circuits and always remember to switch off the supply and disconnect the wiring to your experiment when it is completed.

The procedure for writing reports can be found in Appendix 2 of the author's *Questions and Answers* book. You should write your report in the following sequence, under the subheadings: object, apparatus, diagrams, method, results and conclusions. You should use a 30 cm ruler when drawing circuit diagrams and use a radius aid for drawing circles for instruments. Circuit diagrams are best drawn in pencil (in case mistakes are made) and they should be fully labelled, identifying all the equipment connected in circuit.

When writing your method, do not copy the instruction sheet as this is only for guidance. You should write your method in the past tense. Do not use personal pronouns such as 'I', 'we' or 'our', etc.

Your results will often be a mix of observed results and calculated results. The former being taken from instrument readings while the latter determined from established formulae. Try to tabulate your results and avoid showing too many calculations. If you draw a graph or phasor diagram to illustrate your findings, make sure it is given a title and is drawn to a scale.

Finally, your conclusions. You should look at your results table, calculations, graph and other diagrams and make relevant comments. Refer to the object of the experiment and see if it has been achieved. Sometimes the instruction sheet will ask you to look for further information elsewhere, perhaps your notes or a textbook. Remember, the purpose of doing laboratory work is to give you the opportunity to understand some aspect of the course syllabus whether this be to verify some law or principle or to give you hands-on experience by connecting instruments in circuits or testing equipment.

The first experiment in this chapter is very similar to Example 4.10 found in *Practice 1* book. It investigates star and delta connections and expects you to compare the current and voltage relationships in the two connections when the circuits are balanced and unbalanced. You have an opportunity to find the neutral current in the unbalanced star circuit using a phasor diagram and this should confirm the reading taken by the ammeter connected in the neutral conductor. The results of this experiment are extracted from a student's laboratory book.

4.1 Star-delta connections

Object

To investigate voltage and current quantities in three-phase star and delta connected circuits containing resistance and to determine the neutral current in an unbalanced star connected load.

Apparatus

2 × 0–150 V MI voltmeters
6 × 0–5 A MI ammeters
1 × clamp-on ammeter
3 × 115 W 5 A variable resistors
3 × 0–75 W wattmeters with current coil rated at 5 A and voltage coils rated at 64 V/110 V.

Instructions

1. Make sure that all the instruments are connected and set correctly and that your connection leads are in good condition.
2. Connect the star load as shown in Figure 4.1 and adjust each variable resistor to its maxi-

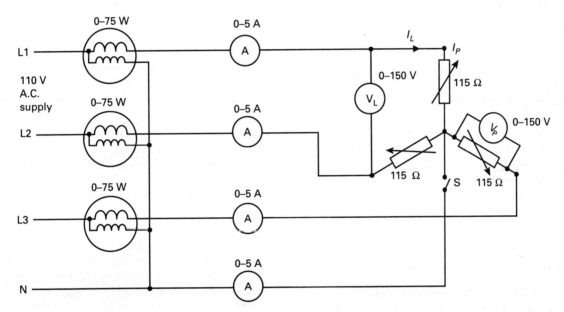

Figure 4.1 Star-connected load

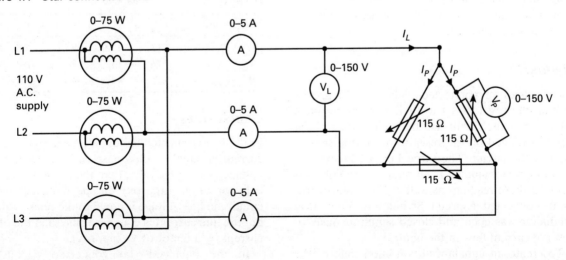

Figure 4.2 Delta-connected load

mum ohmic value. Have the circuit checked before switching on the supply.

3 With the supply switched on, adjust the resistors so that each ammeter reads the same value of current. Do not exceed the current values stated on the resistors. Record all instrument readings.
4 Create an out-of-balance condition by adjusting one of the resistors. Observe the effect and record all instrument readings.
5 Read instruction leaflet on using the clamp-on ammeter and demonstrate its use in the circuit.
6 Switch off the supply and reconnect the circuit as shown in Figure 4.2. Note the different connections and then adjust each resistor to its maximum ohmic value.
7 Have the circuit checked, switch on the supply and re-adjust the resistors to obtain a balanced condition. Note and record all the instrument readings.
8 Switch off the supply and disconnect the circuit wiring.
9 In your report, state the relationship between voltage and current in terms of the line and phase values.
10 Draw a three-phase, phasor diagram of the unbalanced star connected load and find the neutral current. Compare this value with the neutral ammeter reading.

Report

You should copy the object, apparatus list and circuit diagrams as shown on the instruction sheet. Your understanding of the experiment will be judged by the way in which you write method, results and conclusions.

Method

The star load was connected as shown in Figure 4.1 with all the resistors set to their maximum ohmic value. After the circuit had been checked and the supply switched on the resistors were adjusted to balance the circuit. The three line ammeters and the three phase ammeters all read 1.5 A. Note was made of these readings as well as the other instruments connected in circuit. Switch S in the neutral conductor was open and closed simultaneously to observe current flow in the neutral.

To create an unbalanced condition, one of the resistors was adjusted and S was opened and closed. All instrument readings were recorded. To complete this part of the experiment a clamp-on ammeter was used to check the line and phase currents.

The circuit was then reconnected as shown in Figure 4.2 and after being checked and switched on, balance and unbalanced conditions obtained. All instrument readings were noted.

Results

Star connection – balanced

Phase	P	V_L	V_P	I_L	I_P
R	100	113	65	1.5	1.5
Y	102	114	65	1.5	1.5
B	101	115	65	1.5	1.5
N	—	—	—	0.0	0.0

Star connection – unbalanced

Phase	P	V_L	V_P	I_L	I_P
R	100	113	65	1.5	1.5
Y	103	114	65	1.55	1.55
B	200	115	65	3.0	3.0
N	—	—	—	1.45	—

Delta connection – balanced

Phase	P	V_L	V_P	I_L	I_P
R	170	113	112	2.6	1.5
Y	169	114	113	2.55	1.5
B	167	115	112	2.65	1.5

Delta connection – unbalanced

Phase	P	V_L	V_P	I_L	I_P
R	170	113	112	2.6	1.5
Y	260	114	112	3.8	2.3
B	270	115	112	4.0	2.4

Conclusions

In the star connected circuit, it is seen that the line current is equal to the phase current but the line voltage is a higher value than the phase voltage by a factor of √3 (approx). In the delta connected circuit the line voltage is equal to the phase voltage but the line current is a higher value than the phase current by a factor of √3 (approx).

In the balanced star connected circuit, no neutral current flowed but by making the circuit unbalanced, current was able to flow. It was observed that the line and phase ammeters connected to the unadjusted resistors did not alter while the circuit was unbalanced. Figure 4.3 is a phasor diagram showing how the neutral current can be found in the star connected circuit. This compared favourably with the neutral ammeter reading.

The delta connection gives a higher voltage across the load than the star connection and this accounts for the wattmeters giving higher readings. The power demanded in the delta unbalanced condition is much more than its balanced condition.

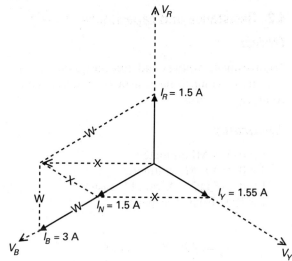

Figure 4.3 Phasor diagram to find neutral current in unbalanced star-connected load

4.2 Resistance and capacitance in series

Object

To investigate voltage and current quantities in an a.c. circuit comprising resistance and capacitance connected in series.

Apparatus

1 × 0–5 A MI ammeter
3 × 0–150 V MI voltmeter
1 × 0–115 Ω/2.8 A variable resistor
1 × 100 µF 110 V capacitor

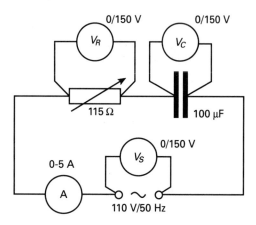

Figure 4.4 RC series circuit

Method

The circuit was connected as shown above in Figure 4.4. The variable resistor was set to its maximum. After checking the circuit connections the supply was switched on. The resistor was then adjusted so that the ammeter read 1 A. Voltmeter readings were taken across the supply terminals and across the two circuit components. This procedure was repeated for adjusted current values of 2 A and 3 A. The observed and calculated results are shown below.

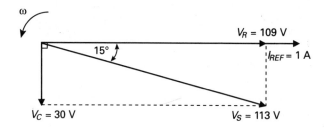

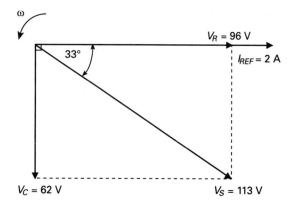

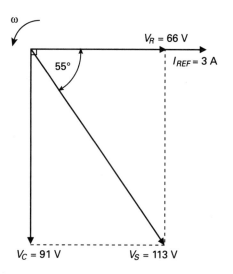

Figure 4.5 Phasor diagrams for a.c. circuit

Results

Observed

V_S	V_R	V_C	I_S	C
113	109	30	1	100
113	96	62	2	100
113	66	91	3	100

Calculated

X_C	R	Z	PF	Ø
30	109	113	0.96	15°
31	48	57	0.84	33°
30	22	38	0.58	55°

Where:

- V_S is the supply voltage
- V_R is the p.d. across the resistor
- V_C is the p.d. across the capacitor
- I_S is the supply current
- X_C is the reactance of the capacitor
- R is the resistance of the resistor
- Z is the impedance of the circuit
- PF is the power factor of the circuit
- $\emptyset$ is the phase angle between supply voltage and current.

Conclusions

The observed results show that the supply voltage remains constant with the p.d.'s across components changing. As more current flows in the circuit, V_R decreases whereas V_C increases. It is seen from the calculated results and the phasor diagrams drawn in Figure 4.5 that as more resistance is taken out of circuit the power factor falls, resulting in a wider phase angle between supply current and supply voltage. The phasor diagram shows the circuit taking a leading power factor.

4.3 Resistance and inductance in series

Object

To investigate voltage and current quantities in an a.c. circuit comprising resistance and inductance connected in series.

Apparatus

1 × 0–5 A MI ammeter
3 × 0–150 V MI voltmeters
1 × 0–115 Ω/2.8 A variable resistor
1 × 0.09 H/3.26 Ω inductor

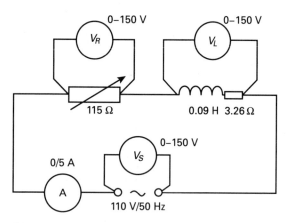

Figure 4.6 RL series circuit

Method

The circuit was connected as shown in Figure 4.6 and the variable resistor adjusted to its maximum full-in position.

After checking the circuit connections the supply was switched on and the resistor adjusted for a current of 1.5 A. Readings on the three voltmeters were taken and recorded. This procedure was repeated for a circuit current of 2.5 A. The observed and calculated results are shown below.

Results

Observed				Calculated				
V_S	V_R	V_L	I_S	X_L	R	Z	PF	$\varnothing$
113	98.5	45	1.5	28.3	65.7	76.7	0.91	24°
113	76	72.5	2.5	28.3	30.4	45.2	0.74	42°

Where:
V_S is the supply voltage
V_R is the p.d. across the resistor
V_L is the p.d. across the inductor
I_S is the supply current
X_L is the reactance of the inductor
R is the resistance of the resistor
Z is the total impedance of the circuit
PF is the power factor of the circuit
$\varnothing$ is the phase angle between supply voltage and current.

Conclusions

It was noticeable that the p.d. across the resistor was a higher value than across the inductor and that by adjusting R more current was taken in the circuit. This caused the power factor to decrease, making the phase angle between current and voltage worse. The calculated values of impedance and power factor are determined using the observed results and recognising that the inductor

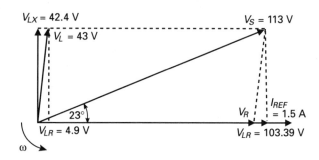

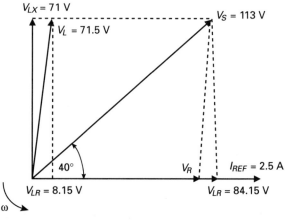

Figure 4.7 Phasor diagrams for a.c. circuit comprising resistance and inductance in series

possessed a small amount of resistance. Figure 4.7 shows phasor diagrams for the two conditions where it can be seen that the supply current lags behind the supply voltage. It is noted that the voltmeter readings (V_L) are not quite the same as found on the phasor diagrams and may be caused by not correctly setting the voltmeter's pointer to zero.

Note: The calculation made to find the power factor and phase angle will not be accurate due to the inductor possessing resistance.

4.4 Resistance, inductance and capacitance in series

Object

To investigate voltage and current quantities in an a.c. circuit comprising resistance, inductance and capacitance connected in series. Also, to determine the circuit's power factor and resonant frequency.

Apparatus

- 1 × 0–5 A MI ammeter
- 4 × 0–150 V MI voltmeter
- 1 × 0–115 Ω/2.8 A variable resistor
- 1 × 50 µF 250 V capacitor
- 1 × 3.2 Ω/0.09 H inductive coil

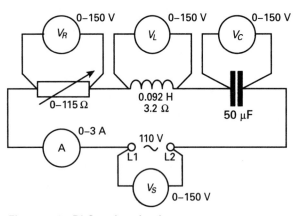

Figure 4.8 RLC series circuit

Method

The circuit was connected as shown above in Figure 4.8. The variable resistor was set to its maximum. After checking the circuit connections the supply was switched on. The resistor was then adjusted to allow 1 A to flow. Voltmeter readings were taken across the supply terminals and across the three circuit components. This procedure was repeated for current values of 1.5 A, 2 A and 2.5 A. The observed and calculated results are shown below together with a calculation of resonant frequency and phasor diagrams for each set of conditions.

Results

Observed					Calculated				
V_R	V_L	V_C	V_S	I_S	R	X_L	X_C	p.f.	∅
104	30	68	115	1.0	104	30.0	68.0	0.90	25.3
94	43	98	115	1.5	63	28.6	65.3	0.82	35.2
78	57	131	115	2.0	39	28.5	65.5	0.68	47.3
55	71	170	115	2.5	22	28.4	68.0	0.48	61.4

Calculation of resonant frequency. Resonant frequency occurs when $X_L = X_C$.

Since:
$$X_L = 2\pi f L$$
and
$$X_C = \frac{1}{2\pi f C}$$
then
$$f_r = \frac{1}{(2\pi \sqrt{LC})}$$
$$= 1/(2 \times \pi \times \sqrt{0.092 \times 50 \times 10^{-6}})$$
$$= 74.3 \text{ Hz}$$

Conclusions

It is noticed that the p.d. across the resistor decreases as it is adjusted to allow more current to flow in the circuit. The increased current increases the p.d.'s across the inductor and capacitor and in the latter case, it is higher than the supply voltage. The values of R, X_L and X_C are calculated from the voltmeter and ammeter readings and are slightly different to the calculations made from the stated values on the components (e.g. the inductor's reactance is 28.9 Ω and capacitor's reactance is 63.66 Ω). Throughout the experiment these values remain unchanged and as more resistance is taken out of the circuit, the power factor decreases and the phase angle between supply current and supply voltage increases (see phasor diagrams shown in Figures 4.9 to 4.12). Since the capacitor has more reactance than the inductor, the circuit takes a leading power factor. The calculation of resonant frequency at 74.2 Hz is a condition whereby X_L equals X_C and if this was achieved the power factor of the circuit would become unity.

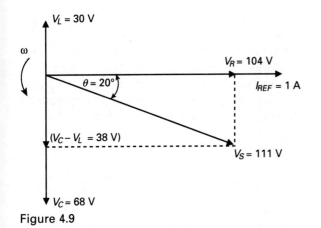

Figure 4.9

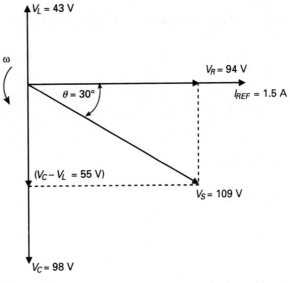

Figure 4.10

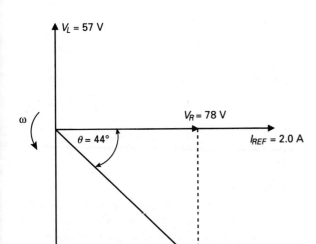

Figure 4.11

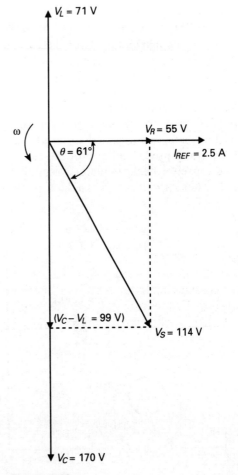

Figure 4.12

4.5 RLC components in series and parallel

Object

To investigate voltage and current quantities in an a.c. circuit comprising resistance and inductance in series, connected in parallel with capacitance. Also, to determine the value of capacitance to achieve unity power factor.

Apparatus

3×0–5 A MI ammeter
1×0–150 V MI voltmeters
1×0–115 Ω/2.8 A variable resistor
1×0–100 µF 110 V capacitor bank
1×6 Ω/0.2 H inductive coil

Method

The circuit was connected as shown in Figure 4.13 with the capacitor bank switched off and the variable resistor set to its maximum. After the circuit connections were checked and the supply switched on the variable resistor was adjusted to allow a current of 1 A to flow in the circuit. Readings were taken of the supply voltage and supply current as well as the p.d.'s across the resistor and inductor.

The capacitor bank was adjusted to 20 µF and when switch S was closed, readings on the supply ammeter and capacitor bank ammeter were taken and noted. Further capacitance was added to the circuit at 40 µF and 60 µF and instrument readings recorded for each condition.

A table of results is given below together with a calculation showing the capacitance required to raise the power factor to unity. Figure 4.14 shows a phasor diagram of the circuit and the effects capacitance has on the RL series components.

Results

Observed							Calculated	
C	V_S	V_R	V_L	I_{RL}	I_C	I_S	p.f.	$\emptyset$
0	120	57	100	1.5	0.0	1.5	0.55	56.6°
20	120	57	100	1.5	0.7	1.0	0.83	33.9°
40	120	57	100	1.5	1.5	0.9	0.94	22.7°
60	120	57	100	1.5	2.6	1.4	0.59	53.6°

Calculation of capacitance for unity power factor.

$R\ = V_R/I_R\ = 57/1.5\ = 38\ \Omega$ (really 44 Ω)
$X_L = V_L/V_R = 100/1.5 = 66.7\ \Omega$
$Z\ = V_S/I_S\ = 120/1.5 = 80\ \Omega$
p.f. $= R/Z\ = 44/80\ = 0.55$ lagging
Hence $\emptyset = 56.6°$

It is noted that the inductor has a resistance of 6 Ω and this is added to the value of R to find the power factor. The diagram in Figure 4.14 shows the inductor's reactive current which needs to be neutralised by the capacitor's reactive current in order to bring the circuit's power factor up to unity.

Since:
$\sin \emptyset = I_L/I_S$
then
$I_L\ = \sin \emptyset \times I_S$
$\quad = \sin 56.6° \times 1.5 = 1.25$ A

This is also the current for I_C. The value of capacitive reactance is found from formula:

$X_C\ = V_S/I_C$
$\quad = 120/1.25 = 96\ \Omega$

Hence
$X_C = \dfrac{1}{2\pi f C}$

and
$C = \dfrac{1}{2\pi f X}$
$\quad = 10^6/(314.2 \times 96)\quad = 33$ µF

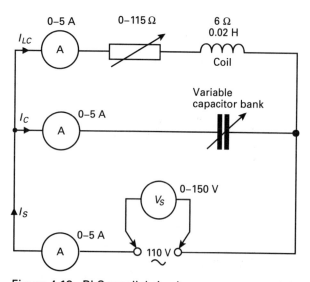

Figure 4.13 RLC parallel circuit

Conclusions

The observed results show no changes in voltage across the supply or across the circuit components. However, it is noticed that changes in the supply current occur, falling and rising to the original value. The power factor of the circuit was found by calculation to be 0.55 lagging before switching in the capacitor bank. This improved to 0.84 lagging when 20 µF was injected in the circuit. Further increases in capacitance at 40 µF and 60 µF made the power factor lead. The phasor diagram in Figure 4.14 illustrates the four conditions, showing that too much capacitance is uneconomical. The calculation made above shows that a 33 µF capacitor is ideal to bring the power factor of the circuit to unity and this would cause the supply current to fall to 0.83 A.

Note: In constructing your phasor diagram, make the scale 1 cm = 0.2 A.

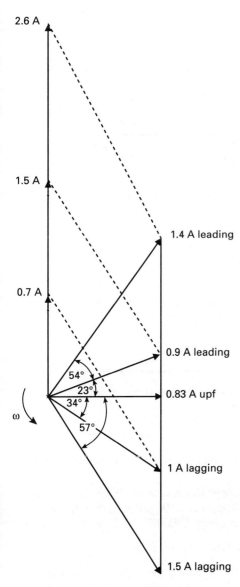

Figure 4.14 Phasor diagram of currents in the capacitor bank

4.6 SOX discharge lamp

Object

To investigate the characteristics and power factor of a low pressure sodium vapour discharge lamp.

Apparatus

1 × 240 V/55 W SOX lamp (inclusive of controlgear)
1 × 240 V/120 W wattmeters
2 × 'Avo' multimeters

Method

The circuit was arranged as shown in Figure 4.15 and the wiring checked. With the capacitor not connected, the supply was switched on. Observations were made of the lamp's colour appearance over a period of ten minutes. The wattmeter and ammeter readings were recorded. The capacitor was switched in circuit and readings of the instruments were taken again.

During full brightness the lamp was switched on and off to see if it would restrike. From a lamp manufacturer's catalogue the lighting design lumens (LDL) of the lamp was found and this was

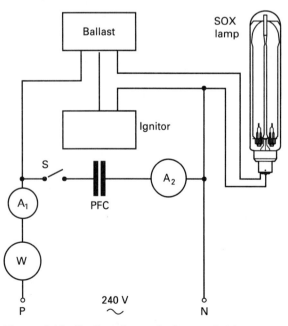

Figure 4.15 Sodium lamp discharge circuit

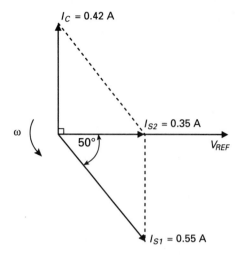

Figure 4.16 Phasor diagram for SOX lamp

used to find its efficacy. Figure 4.16 shows a phasor diagram of the circuit in order to compare the two power factor conditions.

Results

W	A_1	A_2	V	pf	Ø
84	0.55	0	240	0.64	50
84	0.35	0.42	240	1	0

Calculation of lamp efficacy. Efficacy = lumens/watt.

From a lamp catalogue a 55 W SOX lamp has a lumen output of 7,500 after 2,000 hours. Its efficacy is therefore 7,500/55 = 136 lm/W

Calculation of circuit power factor.

$$\text{Power factor} = \text{true power/apparent power} = \text{watts/voltamperes}$$

With no capacitor in circuit:

$$p.f. = 84/(240 \times 0.55) = 0.64 \text{ lagging}$$

With capacitor in circuit:

$$p.f. = 84/(240 \times 0.35) = 1 \text{ or unity}$$

Calculation of capacitor required for the circuit. Capacitive reactance:

$$X_C = V/I = 240/0.55 = 436.36 \, \Omega$$

Capacitor required:

$$\begin{aligned} C &= 10^6/2\pi f X_C \\ &= 1,000,000/(314.2 \times 436.36) \\ &= 7.29 \, \mu F \end{aligned}$$

Conclusions

When the lamp was first switched on it produced a red glow which soon transformed into a monochromatic yellow glow. This changeover is due to the lamp's initial discharge through a mixture of neon and argon gas. The sodium metal gradually vapourises by the heat from the discharge. It was noticed that the lamp operated immediately when it was switched off and then on again.

When the controlgear of the lamp is considered the efficacy value falls to 65% but this is relatively high in comparison to other lamps. Unfortunately, the lamp's colour rendering property is poor, limiting its application to roadways and security lighting. It is seen from the calculation and phasor diagram that the power factor of the circuit is poor and it needs p.f. correction. A further calculation showed that a 7.29 µF capacitor would create unity power factor conditions and this was confirmed by noting that the controlgear was actually wired with an 8 µF capacitor.

4.7 Shunts and multipliers

Note: The instructions for this experiment can be found in the author's *Part 1 Studies: Practical* book.

Object

To investigate the use of shunts and multipliers in the range extension of a universal moving coil instrument.

Apparatus

1 × Model 8 'Avometer'
1 × digital voltmeter
1 × 0–15 mA/75 V universal moving coil meter
1 × Unilab supply unit
1 × 0.1–10 Ω decade resistance box
1 × 315 Ω variable resistor
1 × 39 Ω variable resistor

Method

The circuit was connected as shown in Figure 4.17(a) and the decade box (shunt resistor) adjusted to 0.5 Ω. The supply was then switched on and the variable register adjusted until the multimeter read 20 mA. The decade box was then readjusted until the universal instrument read full scale deflection. The value of decade box resistance was noted. This procedure was repeated for supply currents of 30 mA, 50 mA and 75 mA.

The circuit was then reconnected as shown in Figure 4.17(b) and the decade box (multiplier resistor) was adjusted to 150 Ω. The supply was switched on and the 39 Ω variable resistor adjusted until the digital voltmeter read 0.5 V. The decade box was then readjusted so that the universal instrument read full scale deflection. The value of decade box resistance was noted. This procedure was repeated for supplies 1.0 V, 1.5 V and 2.0 V. A table of results is shown below.

Results

	Shunt Resistor		Multiplier Resistor	
Observed	Calculated	Observed		Calculated
I (mA)	R_S	R_S	V	R_M	R_M
20	15.9	15	0.5	30	28.3
30	4.9	5	1.0	66	61.7
50	2.1	2.1	1.5	102	95.0
75	1.2	1.2	2.0	135	128.3

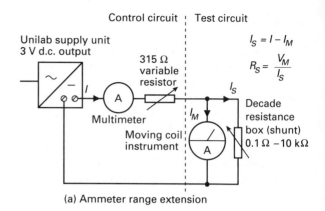

(a) Ammeter range extension

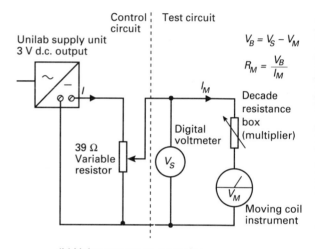

(b) Voltmeter range extension

Where:
R_S is the shunt resistance, read from the decade box
V_M is the maximum voltage, read from the moving coil meter (75 mV)
I is the supply current
I_M is the maximum current, read from the moving coil meter (15 mA)
R_M is the multiplier resistance found on the decade box
V_S is the digital voltmeter supply volts
V_B is the p.d. across the decade box
I_S is the current through the decade box

Figure 4.17 Using a universal instrument as an ammeter and voltmeter

Sample calculations

$R_{SHUNT} = V_M/(I_S - I_M) = 75/(75 - 15) = 1.25 \, \Omega$
$R_{MULTIPLIER} = V_B/I_M = (0.5 - 0.075)/0.015 = 28.3 \, \Omega$

Conclusions

The results table shows little difference between the observed results and the calculated results. The decade box acts as a shunt resistor when connected in parallel with the universal instrument, diverting the supply current away from it. Shunt resistors are used for ammeters. The more current being measured the lower the ohmic value must be of the shunt resistor. Figure 4.17(b) shows the decade box as a multiplier resistor when connected in series with the universal instrument. It is used as a voltmeter.

4.8 Frequency variation

Object

To investigate the effects of variable frequencies on RLC components.

Apparatus

- 1 × frequency oscillator
- 1 × RLC test component board
- 2 × model 8 'Avometers'

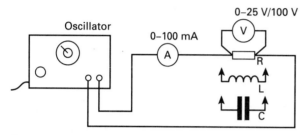

Figure 4.18 Variable frequency supply

Method

The circuit was connected as shown in Figure 4.18. With one Avometer used as a voltmeter, set to 25 V and the other Avometer used as an ammeter, set to 100 mA, the variable frequency oscillator was switched on and allowed to warm up. It was adjusted for a frequency range between 400 and 4,000 Hz. Readings of current and voltage were taken every 400 Hz. This procedure was repeated for the inductor and capacitor with the exception that the voltmeter had to be switched to a higher scale when the inductor was being used. Observed and calculated results are shown below.

Results

	Observed						Calculated (kΩ)		
F	V_R	V_L	V_C	I_R	I_L	I_C	R	X_L	X_C
400	21.5	14.2	24.0	30	65	05	0.72	0.22	4.80
800	21.8	19.8	24.0	30	45	12	0.73	0.44	2.00
1200	21.8	23.8	24.0	30	38	18	0.73	0.63	1.33
1600	21.8	25.0	23.8	30	30	24	0.73	0.83	0.99
2000	21.8	26.2	23.8	30	25	30	0.73	1.05	0.79
2400	21.8	27.8	19.5	30	22	30	0.73	1.25	0.65
2800	21.8	29.7	18.2	30	20	33	0.73	1.49	0.55
3200	21.8	30.0	19.2	30	18	40	0.73	1.67	0.48
3600	21.8	26.8	17.8	30	14	41	0.73	1.91	0.43
4000	21.8	27.9	16.2	30	13	42	0.73	2.15	0.39

Sample calculations. At a frequency of 2,000 Hz:

$R = V/I = 21.8/0.03 = 727 \, \Omega$
$X_L = V/I = 26.2/0.025 = 1048 \, \Omega$
$L = \dfrac{X_L}{2\pi f}$
$ = 1048/(2 \times 3.142 \times 2{,}000) = 0.083 \, \text{H}$
$X_C = V/I = 23.8/0.03 = 793.3 \, \Omega$
$C = 10^6/2\pi f X_C$
$ = 1{,}000{,}000/(2 \times 3.142 \times 2{,}000 \times 793.3) = 0.1 \, \mu\text{F}$
$f_r = 1/2\pi\sqrt{LC} = 1/(6.283 \times \sqrt{0.083 \times 0.1 \times 10^{-6}})$
$ = 1744 \, \text{Hz}$

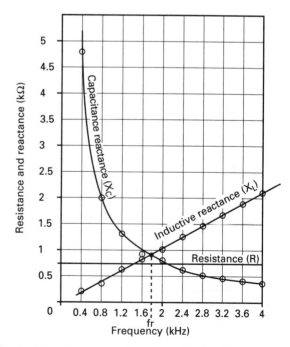

Figure 4.19 Graphs to show the effects of frequency on circuit components

Conclusions

It can be seen from the observed results that changes in supply frequency had no effect on the resistor's ohmic value: it kept constant throughout the experiment. This was not the case for the other two components. At the low end of the frequency range, current in the inductor started off at a relatively high value while current in the capacitor started off at a relatively low value. Figure 4.19 shows graphs of the three components. It is seen that the inductive reactance is proportional to frequency and the capacitive reactance is inversely proportional to frequency. The cross-over or intersection between X_L and X_C occurs at resonant frequency and is shown in the above calculation to be 1744 Hz.

Student activity

Now that you have seen how experiments should be written, see if you can complete the following experiments yourself.

4.9 Universal motor speed–torque characteristics

Object

To investigate the speed–torque characteristics of a universal motor using a brake test.

Apparatus

1 × single-phase universal motor 250 W/100 V
1 × 0–30 A MI ammeter
3 × 0–150 V MI voltmeter
1 × 0–300 W/120 V wattmeter
1 × brake test stand
1 × portable stroboscope

Student activity

1. Connect the circuit as shown in Figure 4.20 and have the connections checked.
2. Measure the radius (r) of the motor's pulley in metres and make sure the pulley brake belt is correctly fitted around the pulley before switching on the supply.
3. Check the motor's rated speed and switch on the portable stroboscope to the desired range.
4. Start the motor running and tighten the belt on the brake to a point where the motor nearly stalls.
5. Compare the ammeter reading with the motor's rated current.
6. Note the effective pull (F) at the circumference of the pulley, taking the difference between the spring balance readings. Record the speed (n).
7. Take at least six other readings, gradually allowing the motor to run faster but not out of control.
8. For each speed (rev/s), determine the motor's torque from the formula $T = F \times r$ Nm.
9. Tabulate your results and plot a graph of the motor's speed–torque characteristics.
10. Write a laboratory report.

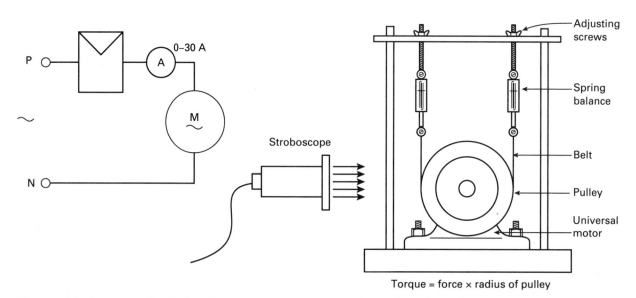

Figure 4.20 Apparatus for finding the speed–torque characteristics of a universal motor

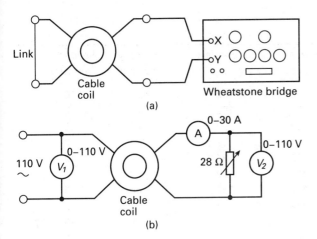

Figure 4.21 Voltage drop in a cable
(a) Measurement of resistance
(b) Determining voltage drop

4.10 Voltage drop in a cable

Object

To investigate several methods of finding voltage drop in a 50 m length of 1.5 mm² PVC/PVC/CPC (twin and earth) cable.

Student activity

Find the cable resistance using an ohmmeter or Wheatstone Bridge (see Figure 4.21(a)). Assume it to be 2.293 Ω.

Now find the cable's resistance using Table 9A of your *IEE Site Guide* and the formula:

$$R = (\rho L/A)$$

where $\rho = 17.2$ μΩ mm, $L = 200$ m and $A = 1.5$ mm². Compare both values.

Connect the circuit as shown in Figure 4.21(b).

Test 1 – with full resistance in circuit, switch on and adjust the variable resistor to allow 5 A to flow. Take readings of both voltmeters V_1 and V_2.

Test 2 – adjust the variable resistor so that the ammeter reads 10 A. Take readings of both voltmeters.

Test 3 – repeat above for ammeter reading of 15 A.

a) Find the voltage drop in the cable using the formula:

$$V_C = I \times R \quad (1)$$

b) Find the voltage drop by subtracting V_2 from V_1:

$$V_M = V_1 - V_2 \quad (2)$$

c) Find the voltage drop by using Table 6D2 of the *IEE Site Guide* and the formula:

$$V_R = L \times I_b \times \text{mV/A/m} \quad (3)$$

Complete the following results table and write a laboratory report about the different methods of finding voltage drop.

Results

| Test | Observed | | | Calculated | | |
| | | | | Measured | Calculated | Regs |
	I	V_1	V_2	V_M	V_C	V_R
1	5	29	15			
2	10	56	31			
3	15	84	46			

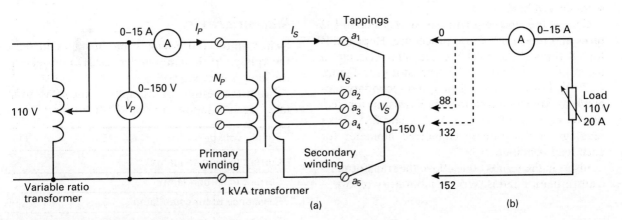

Figure 4.22 Transformer test circuit

Results

V_p V_s	N_p N_s	I_s I_p	Primary voltage	Secondary voltage	Primary turns	Secondary turns	Primary current	Secondary current
1:1	1:1	1:1	30	30	152	152	10	10
			50	50	152	152	8	8
			80	80	152	152	5	5
			100	100	152	152	2	2
			30	26	152	132	8.7	10
			50	43.5	152	132	6.9	8
			80	69	152	132	4.5	5
			100	87	152	132	1.7	2
			30	17.5	152	88	5.8	10
			50	29	152	88	4.6	8
			80	46	152	88	3.0	5
			100	58	152	88	1.2	2

4.11 Transformer ratios

Object

To verify that the ratio of a transformer's primary to secondary voltages on open circuit is equal to the ratio of its primary to secondary turns, and also, that both these ratios are approximately equal to the ratio of its secondary to primary current.

Student activity

With the circuit connected as shown in Figure 4.22(a) and the secondary tappings equal to the primary tappings (i.e. 152 turns each), switch on the supply Variac and adjust it to deliver 30 V, 50 V, 80 V and 100 V to the primary winding. For each of these voltages, take the readings on the secondary voltmeter.

Change the tappings on the secondary side to 132 turns and repeat the above procedure. Repeat this for 88 turns. Reconnect the secondary circuit to incorporate the variable load resistor (see Figure 4.22(b)). For the same primary and secondary tappings as above, adjust the load resistor for secondary currents of 10 A, 8 A, 5 A and 2 A. Take readings of the primary connected ammeter for each load condition.

Insert in the results table above the transformer's transformation ratios. Write a laboratory report.

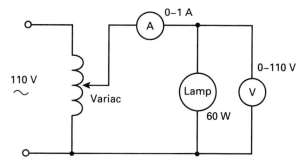

Figure 4.23 Lamp circuit

4.12 Temperature coefficient of resistance

Object

To show how the resistance of tungsten and carbon filament lamps vary with temperature (Figure 4.23).

Student activity

Switch on the 110 V supply and in stages, increase the voltage to the lamp under test. Do not exceed the lamp's working voltage.

From the results, plot graphs of the lamp resistance against supply current. Write a report.

Supply voltage	15 $\longrightarrow$	110
Current in tungsten lamp		
Current in carbon lamp		
Resistance of tungsten lamp		
Resistance of carbon lamp		

4.13 Three-phase induction motor

Object

To investigate the running characteristics of a three-phase, cage induction motor.

Student activity

The experiment is shown in Figure 4.24. Before switching on the supply, familiarise yourself with the brake test equipment and the operation of the portable stroboscope. The motor's nameplate will tell you the rated current and rotor speed. If required, fill the motor's pulley with water and start the motor running with the pulley belt loose.

Record the motor's speed on no-load, taking readings of the connected instruments.

See Experiment 4.9 on using the brake and make several tests with one at full load.

Create a table of results from your instrument observations. Make calculations of output power, input power, power factor, per unit slip and efficiency. Write a laboratory report.

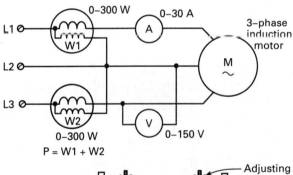

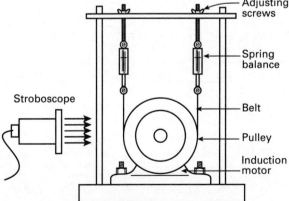

Figure 4.24 Brake test for a 3–phase, cage rotor induction motor

Results

Observed			
Input power (W)	Supply voltage (V)	Supply current (A)	Speed (rev/s)

Calculated				
Output power (W)	Torque (N_M)	Power factor (cos ø)	Efficiency (per unit)	Slip (per unit)

4.14 Power factor correction

Object

To observe the effect of connecting various values of capacitance to a switch-start fluorescent lamp circuit.

Student activity

The circuit is shown in Figure 4.25 which uses a variable capacitor bank. Connect the circuit to the supply and adjust the capacitor bank in steps of 0.5 µF. Record the wattmeter, voltmeter and ammeter readings and determine the power factor in each case. Plot a graph of power factor and current against capacitance. Determine from the graph, unity power factor conditions. Write a laboratory report.

Results

Capacitance (µF)	Power (W)	Voltage (V)	Current (A)	Volt-amperes (VA)	Power factor (cos ø)
0		240			
0.5		240			
1.0		240			
1.5		240			

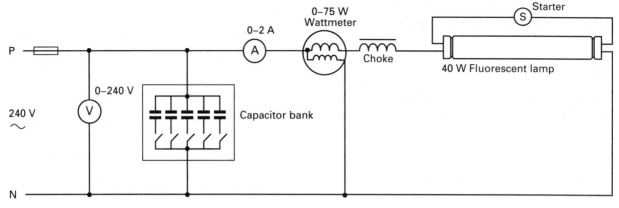

Figure 4.25 Power factor correction in a fluorescent lamp circuit

4.15 Rectifier circuits

Object

To investigate the waveforms of half-wave and full-wave rectifiers.

Student activity

Arrange the rectifier circuits as shown in Figure 4.26. Connect, in turn, each rectifier circuit to the lamp load and observe the waveshape on the oscilloscope. Sketch the fundamental waves and the rectified waveshapes and write a laboratory report.

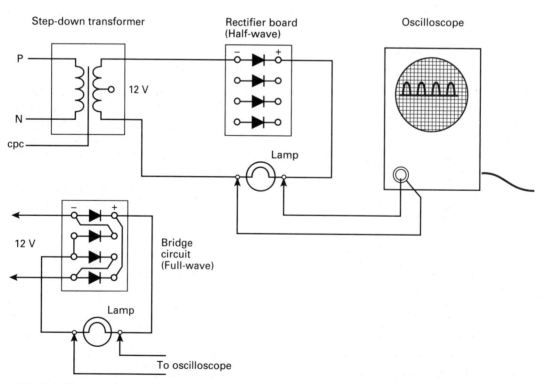

Figure 4.26 Rectifier circuits

4.16 Audio frequency amplifier

Object

To investigate the output waveform of an audio frequency amplifier.

Student activity

Arrange the amplifier circuit similar to Figure 2.28. Connect a low frequency generator and voltmeter across the input of the amplifier and set the generator to give a reading of 10 mV on the voltmeter at a frequency of 5 kHz. Remove the voltmeter from the input and connect it across the output. Read the output voltage and find the voltage gain (output/input).

Connect the oscilloscope across the output of the amplifier and trace the waveform. Raise the input signal to 50 mV and note the effect and again trace the waveform.

Note: On the signal generator select the high impedance output, a frequency band between 1 kHz and 5 kHz and amplitude 10 mV to 50 mV. On the oscilloscope, select a channel, the a.c. switch, the correct amplitude and time base setting.

4.17 Multimeters

Object

To investigate the switching functions of a Model 8 'Avometer' and to measure various circuit quantities.

Student activity

Connect the meter leads to the instrument ensuring that each coloured lead is connected to the same coloured terminal.

Read the working instructions on the back of the instrument or the model's instruction sheet.

Note: To adjust the pointer, set the left-hand switch to resistance and then join the leads together:

On the Ω range, adjust to zero by means of the knob marked ZERO Ω;
On the Ω ÷ 100 range, adjust to zero by means of the knob marked ZERO Ω ÷ 100;
On the Ω × 100 range, adjust to zero by means of the knob marked ZERO Ω × 100.

To test for resistance, set the right-hand switch at the range required and the leads across the unknown component. The resistance will be read on the Ω range or by switching to the other ranges.

For current and voltage measurements see manufacturer's instructions.

4.18 Star-delta starter

Object

To investigate the starting and running performance of a six-terminal cage induction motor connected to a star-delta starter.

Student activity

Arrange the circuit as shown in Figure 4.28. Switch on the supply and start the motor. Estimate the surge current and take measurement of the star running current. Also record the star connected voltage.

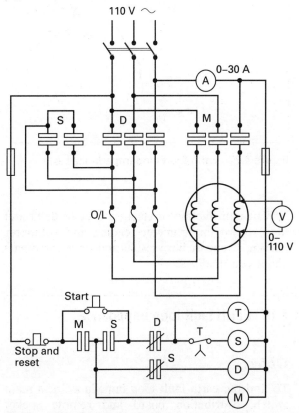

Figure 4.28 Star-delta starter

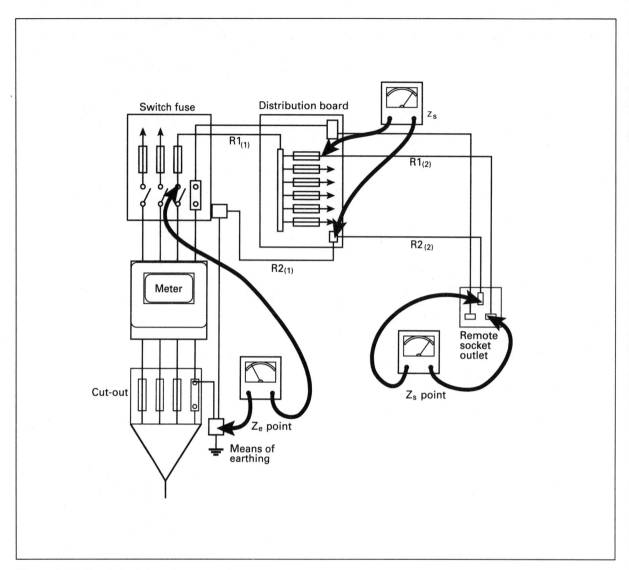

Figure 4.27 Earth fault loop impedance tests

Wait until the motor changes over to delta and then record the ammeter reading and voltmeter reading. Compare both results and write comments about your findings.

4.19 Earth fault loop impedance

Object

To find the earth fault loop impedance at a main switch, distribution board and remote socket outlet.

Student activity

Study Figure 4.27 and read the instructions for the operation of your tester. The instrument must obtain a supply from the same circuit. Carry out the tests as shown in the diagram. Once you have found the values, determine the prospective short circuit current at the origin of the installation, the distribution board and socket outlet. Check your results with the short circuit ratings of the circuit protective devices. Remember that your tests results are made cold and not under fault conditions. State the maximum Z_E values for different earthing systems outside a consumer's installation.

4.20 Transformer tests

Object

To investigate the efficiency of a 1 kVA/110 V primary double-wound transformer.

Student activity

Arrange the circuit as shown in Figure 4.29(a) which allows an open circuit test to be made. The wattmeter reads the core or iron losses. Reconnect the circuit as shown in Figure 4.29(b), which allows a short circuit test to be made. Switch on the supply starting at 0 V and carefully regulate the supply voltage to circulate 10 A in the secondary winding. The wattmeter measures the winding or copper losses. Determine the efficiency of the transformer by adding the two losses together and using the formula:

Efficiency = output/(output + losses)

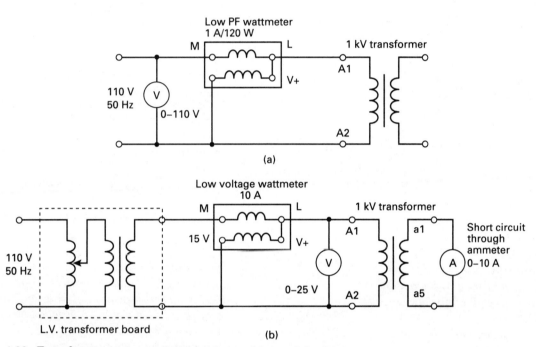

Figure 4.29 Transformer test   (a) Open circuit test (b) closed circuit test

Terminology and further information 5

Objectives

After working through this chapter you should be able to:

- *know the meaning of the following terms:*

    competent person
    clamp-on ammeter
    volt drop
    residual current device
    wiring system
    radial circuit
    cage induction motor
    contactor
    signal generator
    oscilloscope
    logic probe
    frequency meter
    graphical symbols
    overlay
    diversity factor
    lumen method of calculation
    balanced and unbalanced loads
    efficacy
    shunts and multipliers
    universal motor
    transformer ratios
    temperature coefficient of resistance
    power factor
    rectification
    audio amplifier multimeter
    earth fault loop impedance
    star-delta starter

- *know the*:

    quantities an ammeter, voltmeter and wattmeter measure;
    safety precautions when using test equipment
    essential parts of a motor;
    characteristics of different light sources;
    properties of circuits containing RLC components;
    basic features of a universal motor and induction motor;
    tests for determining losses in a transformer;

- *know the requirements for:*

    measuring voltage in lamp circuits;
    inspecting and testing electrical installations;
    testing and measuring electronic equipment;
    sequence of control in an installation;
    emergency switching;
    selecting suitable wiring systems;
    determining the numbers of lamps for a room.

5.1 Installation practices and technology (Chapter 2)

Competent person (Task 2.1). A person who is likely to have a mix of technical knowledge and experience or be so supervised as to be capable of ensuring that danger is prevented. This implies that the person has an adequate understanding of the system or circuit to be worked on, as well as an understanding of hazards which may arise and precautions which need to be taken. The person must also have the ability to recognise at all times whether it is safe for work to continue.

Clamp-on ammeter (Task 2.2). A hand-held portable instrument for measuring current in a circuit without disturbing circuit connections. **Note:** A *digital clamp-on power meter* is a similar type of instrument which is designed to measure voltage, power, apparent power, frequency and power factor. It can also be used with an oscilloscope and chart recorder to provide an analogue output.

HSE Guidance Note GS 38 (Task 2.3). 'Electrical test equipment for use by electricians' advises on the selection of suitable test probes, leads, lamps, voltage-indicating devices and other measuring equipment used by electrically competent people when working on or near circuits not exceeding 650 V. It states the dangers and causes of accidents and it provides information on making test equipment safe. It also provides information on the procedure for a safe system of work.

Voltage drop (Task 2.4). A product of current and resistance. If it exceeds a certain level it causes connected equipment to operate inefficiently. This means the equipment cannot demand the required power for it to function properly. **Note:** Conductor resistance is a relatively low ohmic value which is proportional to its length (mΩ/m). Voltage drop tables in the *IEE Wiring Regulations* state voltage drop in terms of mV/A/m which is the same as mΩ/m.

Measuring instruments (Task 2.5). It is important to remember that an *ammeter* measures current and should be connected in series with the load. A *voltmeter* measures potential differences and should be connected across live conductors. A *wattmeter* measures the power consumed in the circuit and is connected in series with the load (current coil) and across live conductors (voltage coil).

Residual current device (Task 2.6). Previously called an *earth leakage circuit breaker* and a form of circuit protective device only from the point of view that it is connected in a circuit. It operates when an earth fault is detected but not for a short circuit fault. It should not be confused with a *miniature circuit breaker* which, like a fuse, is a true circuit protective device.

Wiring system (Task 2.7). Defined in the 16th Edition of the *IEE Wiring Regulations* as an assembly made up of cable or busbars and parts which secure and, if necessary, enclose the cable or busbars. There are many examples, such as MIMS cable, PVC-armoured cable, PVC single-core cables in conduit or trunking, multicore PVC-insulated cable. See Table 4A of the IEE Wiring Regulations. **Note:** In the example given in this task, the wiring system is a *radial circuit*, taken from its own fuseway in the consumer unit and not from a 13 A socket outlet connected to the ring circuit. It is important to use heat-resistant flexible cord to connect the immersion heater. Use your *IEE Wiring Regulations* and *Site Guide* and know what correction factors to use.

Radial circuit (Task 2.8). Not to be confused with a ring circuit. Radial circuits are lighting circuits, immersion heater and cooker circuits. **Note:** Main distribution systems are often wired as radial or ring main feeders and must not be confused with radial or ring final circuits (which were once called final subcircuits).

Cage induction motor (Task 2.9). This motor was once called a *squirrel cage motor* because of the design of its rotor. The cage needs no electrical connection since its operating principle is by electromagnetic induction. The starting arrangement for a 3-phase motor is either by direct-on-line contactor starter or star-delta starter. Both types incorporate undervoltage protection and overload protection.

Motor parts (Task 2.10). The essential parts of an a.c. motor are its *stator* frame and *rotor*. The former is the stationary part and the latter its rotating part. The stator frame is often of aluminium, cast iron or steel construction. Inside the stator frame will be found a steel lamination core pack which is either pressed or shrunk-fit into the stator frame. It supports the stator windings which are insulated with a synthetic resin. The rotor is a die-cast aluminium (or fabricated copper) cage, fixed to the driving shaft of the machine. It comprises *endrings* and often an integral fan to dissipate heat. The rotor shaft is made of high grade steel and has a keyway. The drive shaft extension is drilled and tapped for removal of the driving pulley or coupling. High quality ball or roller bearings are assembled on the shaft. The motor's endshields are of die-cast aluminium or cast iron construction although fabricated steel is chosen for larger motors. The end shields are designed to give maximum support to the rotor assembly.

Contactor (Task 2.11). Used for the control of large loads and often operating through a *time switch*. The contactor coil is connected through a number of normally open or normally closed switches or stop and start buttons. Where the control circuit is wired from the secondary winding of a step-down transformer or the cross sectional area is reduced, fuses should be inserted in the circuit. **Note:** You should read group Regs. 476–03 and 537–04 of the *IEE Wiring Regulations* concerning devices for emergency switching.

Lighting circuits (Task 2.12). Of the lamp circuits listed in this task, only the fluorescent lamp and three discharge lamps require control gear. You will find the selected GLS and TH lamps produce a white light with excellent colour rendering (the effects that a light source has on the appearance of coloured objects). The TH lamp has a higher *efficacy* (more light for the same power) and a longer *life expectancy*. Care is required when handling the lamp.

In the switch-start MCF lamp, most of the light is produced from a phosphor coating on the inside of the tube. *Colour rendering* depends on the colour appearance of the lamp, varying between good (2), very good (1B) and excellent (1A). The switch-start circuit should re-start instantly and an electronic starter used.

The MBF lamp is a high intensity discharge lamp and has a universal burning position. Its colour rendering varies between acceptable (3) and good (2) and it produces a cool white light. It takes several minutes to *restrike* once it has been switched off.

The SOX lamp is a high intensity discharge lamp and operates at low pressure. Low wattage lamps can restrike instantly after being switched off and can be used in the horizontal or vertical (cap up) operating position. Higher wattage lamps take about ten minutes to restrike and require a horizontal operating position. Unfortunately because of the lamp's monochromatic yellowish light, colour rendering does not occur but it is a very efficient light source.

The SON lamp is a high intensity, high pressure lamp and its burning position is universal. Its restrike time is less than one minute and it produces a golden white light. Colour rendering is poor (4) but it is a very efficient light source. **Note:** Electronic *control gear* is used more and more today for the starting and running operation of fluorescent and discharge lamps.

SON lamp control gear (Task 2.13). There are several versions of the SON lamp and power ranges between 50 W and 1,000 W. A typical Thorn 70 W lamp requires a *power factor correction capacitor* of 10 µF and the lamp takes a *starting current* of 0.55 A and *running current* of 0.4 A. You will discover that the p.f. correction capacitor makes no change to the working of the lamp or its starting. Its use in the circuit is primarily to keep the circuit power factor high and reduce the supply current. The lamp circuit incorporates an *ignitor* which is intended to generate voltage pulses to start the lamp. These pulses are of a short duration and because of the capacitive attenuation (line loss), the length of cable between the ignitor and lamp is limited. In the case of the 70 W SON lamp, the maximum cable capacitance when using a Thorn G53353.4 ignitor is 700 pF In order to find the maximum length of cable to use, the maximum cable capacitance is divided by the capacitance per metre of the cable selected for the wiring. For example, a 1.5 mm^2 flat twin and earth PVC sheathed cable has a capacitance per metre of 115 pF and this allows the ignitor to be placed 6 m away from the lamp.

MCF lamp voltage measurement (Task 2.14). It is very important to exercise care when taking voltage measurements, making sure the instrument is initially set on the highest range. If you use a digital voltmeter you only need to glance at the instrument to note its reading. This allows you to concentrate on holding the test leads properly at the measuring points. You will discover that different voltages appear, for example, across the starter side of the tube and the neutral, the voltage will only be about 3 V whereas across the glow-type starter terminals it will be about 200 V.

Test instruments (Task 2.15). There are various proprietary instruments which can be used take measurement of resistance, current and voltage.

The guidance notes in this task offer the instruments that are most likely to be used for the quantities being tested. For example, insulation resistance must be capable of being read in megohms and the RCD tester must display measurement of time. As described above, exercise care when making the tests.

Testing electrical installations (Task 2.16). See comments made above for Task 2.15.

Testing with electronic equipment (Task 2.17). Digital and analogue multimeters are universal instruments for measuring various electrical quantities such as a.c./d.c. voltages and currents and resistance, capacitance, frequency, etc. See Figure 2.27(a).

A *signal generator* is an oscillator which produces an a.c. voltage of a continuously variable frequency (0.1 Hz to 100 kHz.) One of its applications is to test the frequency response of an audio amplifier over a range of frequencies (see Task 2.18).

An *oscilloscope* is a high impedance voltmeter and displays the voltage applied to its input, showing how it varies with time. A typical dual beam oscilloscope is shown in Figure 5.1. The procedure for displaying a signal is given in the extract below.

> Disconnect the signal leads if fitted.
> Switch on the supply and wait about 20 seconds for the tube to warm up.
> Set the intensity control (1) to approximately half maximum.
> Set bright line control (2) to ON and set mode control (3) to DUAL. This should produce two horizontal lines on the screen. If not, adjust the channel 1 and channel 2 'Y' controls (4) until the lines are approximately central on the screen.
> Adjust the focus control and intensity controls to give the best trace.
> To display a signal, connect one of the signal leads to channel 1 input (5). If there is a waveform on the screen it means the other end of the lead has picked up 'noise' – which can happen when the lead is open circuit. The signal will disappear if the lead ends are shorted together.
> Set channel 1 gain control (6) to a sensitivity of 5 V/cm and also set channel 1 input switch (7) to AC.
> Set the time-base speed control (8) to 0.2 ms/cm (approx vertical position) and also set variable sweep control (9) to the CAL position (fully clockwise).
> Set the time base speed magnification switch to X1 (10).
> Switch on the signal generator and set the frequency to 1 kHz, voltage output control to minimum position and attenuator set to 0 dB.
> Connect the free end of the channel 1 input lead to the output of the oscillator and increase the output voltage until the trace covers approximately 2/3 of the screen height.
> To synchronise the signal with the time base sweep, set the trigger selector (11) to channel 1 and adjust the trigger level control (12) slowly until the trace 'locks'.
> Gradually reduce the output from the signal generator until the scope's trace is in synchronism.
> Set the oscillator output to give a trace of approximately 2/3 screen height with channel 1 gain control remaining at 5 V/cm.
> Adjust the X shift to centralise the trace and set the trigger selector switch to EXT.
> The trace will now become unsynchronised.

A *logic probe* is a small hand-held instrument used for checking the logic levels of integrated circuits (ICs), either TTL (transistor-transistor logic) or CMOS (complementary metal oxide semiconductor logic) circuits. See Figure 2.27(c). The checking is often done on the pins of an IC chip. The probe incorporates light emitting diodes (LEDs). In instruments with two LEDs, logic 0 is indicated by a green LED and logic 1 is indicated by a red LED. If both LEDs are off it indicates an open circuit. If they both flash or are both dim the input signal is a square wave or it consists of repetitive wide pulses. The probe also tests repetitive narrow pulses, both negative and positive over a range of frequencies.

A *frequency meter* is used to measure frequency over specified ranges. It can be designed as a portable hand-held instrument or bench top instrument. See Figure 2.27(d). The hand-held instrument is battery operated and often has a large 8-digit display, capable of measuring frequencies up to 1300 MHz. The instrument has a 'press to measure' and 'hold' facility.

Measurement of amplitude and frequency (Task 2.18). To measure amplitude and frequency, connect a digital voltmeter (DVM) set on the 20 V a.c. range to the output of the signal generator. Set the

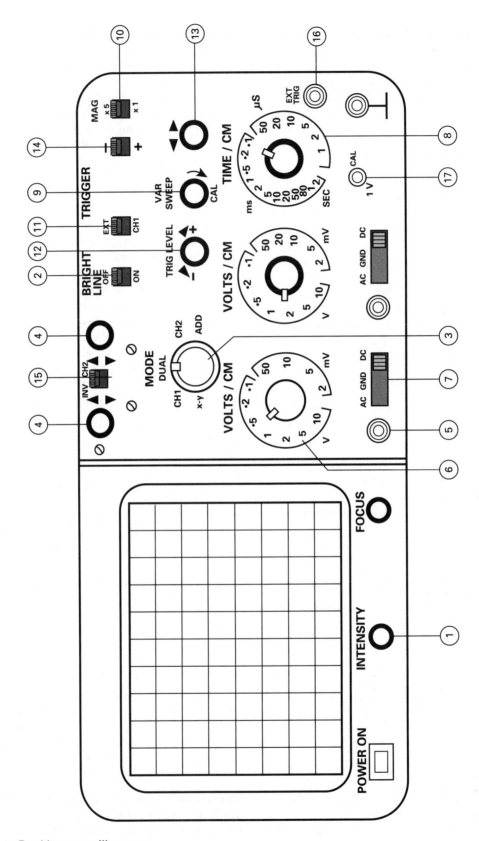

Figure 5.1 Dual beam oscilloscope

frequency to 1 kHz and the signal amplitude to 4 V using the DVM. Set the oscilloscope channel 1 gain control to 2 V/cm, the time base speed control to 0.2 ms/cm and the input condition to AC. Connect the channel 1 input to the signal generator leaving the DVM connected.

To measure *amplitude*, adjust the trace with the Y shift control until the bottom of the trace coincides with a graticule marking. Observe the top of the trace and measure the height in cm. (Use the X shift control to make the peak coincide with the centre scale.) The peak-to-peak amplitude (in volts) will be the number of cm measured times the setting of Y gain control. Compare this reading with the DVM indication of 4 V r.m.s. Repeat the procedure for DVM values of 0.4 V and 0.1 V using appropriate settings of the gain control to give a reasonably large trace on the screen. Use the attenuator on the oscillator for low voltage output.

To measure *frequency*, set the output of the oscillator to approximately 4 V r.m.s. at a frequency of 1 kHz. Select a suitable Y gain setting to give a trace covering approximately 2/3 of the screen height and set the time-base speed control to 0.2 ms/cm. Using the X shift control, measure the length of 1 cycle in cms. (For greater accuracy, measure this on the slope of the waveform rather than on the peaks.) The time for one cycle will be: T = Number of cms measured times the setting of the time-base speed. The frequency is $f = 1/T$ Hz. Check the measured frequency with the setting on the signal generator. Repeat for indicated frequencies of 30 Hz and 25 kHz choosing the appropriate time-base settings to give time for 1 cycle of 1/2 to 2/3 screen width.

Regulation test requirements (Task 2.19). A continuity resistance test is made to check the soundness of circuit conductors. In a ring circuit there should not be any breaks or interconnectors which can by-pass some of the socket outlets. The *insulation resistance test* is to detect any leakage current flowing to earth and for the test to be satisfactory a reading of 0.5 MΩ is required. Polarity of socket outlet connections is important since it might cause connected equipment to be a source of danger especially if disconnected for repair or maintenance.

Lamp circuit faults (Task 2.20). The SON lamp has already received attention in Task 2.13. In tracing the circuit component faults, check for loose connections to the ignitor and also check to see if the wiring insulation is sound, especially between the choke, ignitor and lamp. Check the length of cable between the ignitor and lamp to see if it is within the limits. Check the mains voltage with a proprietary digital voltage tester. You may have to substitute components. Replace the lamp if the voltage drop is above 150 V (for 50 W and 70 W lamps).

5.2 Course assignments (Chapter 3)

Graphical symbols (Example 1). Figure 5.2 shows a selection of BS 3939 graphical symbols for electrical power, telecommunications and electronic diagrams. Figure 5.2 includes a list of the equipment shown by the BS 3939 symbols. (A scale for the drawing could be used to find out how much conduit and cable would be required from which a *material requisition* could be produced.)

Sequence of control (Ex. 2). All electrical installations must be provided with (i) a means of *isolation*, (ii) means of *excess current protection* and (iii) means of *earth leakage protection*. You should read pages 3–18 of the IEE *Site Guide*.

Motor control gear (Ex. 3). Consult product catalogues to find out the different types of switchgear and control gear for this installation. Thermal overloads are provided in the majority of motor starters. Refer to pages 60–61 of the author's *Part 2 Studies: Science* book for more information.

Overlays (Ex. 4). The use of tracing paper over drawings to show wiring routes is not new to Part 2 students. In many instances, however, not much information is given about the routes taken or the sizes of cables or conduits, etc. used. The overlays should be given a legend and show the outline plan of the building and its internal rooms. The routes taken by different wiring systems should be identified. Use coloured pencils. **Note:** Figure 5.2 shows a selection of BS 3939 graphical location symbols. The storage heater symbol in this exercise is not recognised as a BS 3939 location symbol and it must not be mistaken for a heating distribution board.

Emergency switch circuit (Ex. 5). This is to provide the rapid disconnection of the supply to prevent or remove danger. Switches must be readily accessible and marked with a red handle or push button. They must be double-pole and latch operated and be

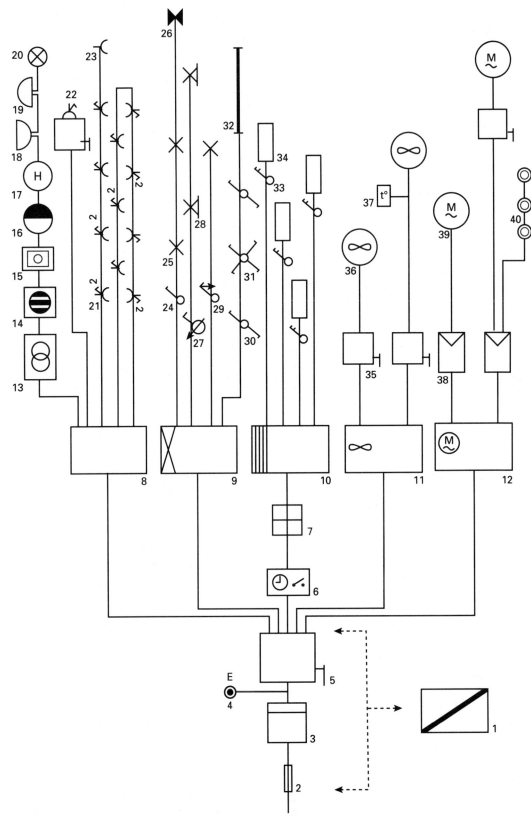

Figure 5.2 BS 3939 graphical symbols

Legend
01 Main control or intake point
02 Fuse (cut-out type)
03 Integrating meter
04 Consumer's earthing terminal
05 Main switch
06 Time switch
07 Contactor
08 Distribution board
09 Lighting distribution board
10 Heating distribution board
11 Ventilating distribution board
12 Motor distribution board
13 Transformer
14 Indicator panel
15 Restricted access fire alarm push button
16 Smoke detector (accepted)
17 Heat detector (accepted)
18 Buzzer
19 Bell
20 Signal lamp
21 Twin switched socket outlet
22 Cooker control point
23 Socket outlet
24 Single-pole one-way switch
25 Lighting point
26 Emergency (safety) lighting point
27 Dimmer switch
28 Wall mounted lighting point
29 Cord-operated, single-pole, one-way switch
30 Two-way switch
31 Intermediate switch
32 Single fluorescent luminaire
33 Two-pole, one-way switch
34 Electrical appliance
35 Main switch
36 Fan
37 Thermostat
38 Starter
39 A.C. motor
40 Push button (emergency)

capable of cutting off the full load current. You will see from circuit B in this task that the stop buttons will activate a contactor coil and you are asked in this exercise to draw the complete circuit. **Note:** It is normal practice in these types of circuit to install a red warning lamp near the main control panel in order to indicate that the circuit has been activated and needs investigating. You should show this on your circuit diagram as an additional item.

Diversity factor (Ex. 6). In this exercise (part (b)) you are asked to perform a calculation to find the current taken by one of the fluorescent luminaires. As with final circuits for discharge lighting, consideration has to be given to the current taken by the luminaire's control gear and any harmonic current which the control gear might possibly generate. Where exact information is not available, the voltamperes of the circuit is the lamp's rated wattage multiplied by a factor of 1.8. This figure is chosen, based on the assumption that the circuit is corrected to a power factor of not less than 0.85 lagging.

Installation practice (Ex. 7). In this exercise you are given the opportunity to select suitable size conduits and trunking for the motor circuits. In practice, it is often a rule-of-thumb approach which is used, based upon experience. Sometimes this decision can go wrong and leads to PVC sheathed cables becoming stressed and overheated in their enclosures. Consideration must be given to *space factor* and *grouping factor* – see Table 4B1, Appendix 4 of the *IEE Wiring Regulations*.

TN-C-S earthing system (Ex. 8). See definition in Part 2 of the *IEE Wiring Regulations* and also page 14 of the IEE *Site Guide*.

Lumen method (Ex. 9). This is a calculation method for determining the number of luminaires needed to provide a given luminance. The basic steps are as follows:

Step 1. Calculate the room index (RI), floor cavity index (CI) and ceiling cavity index (C).
Step 2. Determine effective reflectances of the ceiling, walls and floor cavities.
Step 3. Determine the utilisation factor (UF) for the luminaires at the room index and reflectances in steps 1 and 2 above.
Step 4. Determine any correction factor (CF) for the type of luminaire being installed – see photometric data.
Step 5. Determine the bare lamp flux (F) for the selected lamps.
Step 6. Determine the appropriate light loss factor (LLF) or maintenance factor (MF).
Step 7. Determine the number of luminaires (N) required from the formula:

$$N = \frac{E \times A}{F \times n \times \text{LLF} \times \text{UF} \times \text{CF}}$$

where E is the required average illuminance and A is the area of the working plane.

Note: In Step 1:

$$\text{RI} = (L \times W)/H(L + W)$$

where L is length of room, W is width of the room and H is height of luminaires above the working plane.

In Step 2:
$$CI = RI \times H/H_f$$
where H_f is the height of the working plane.

Once you know the reflectances and room indices, photometric tables will tell you what utilisation factors to use.

To find the ideal ceiling arrangement for the selected luminaires to give uniformity of illuminance, you must first find the maximum spacing to height ratio (SHR). This is measured between the centres of adjacent luminaires. The spacing should not exceed the SHR maximum given in photometric tables. You will come across the terms SHR MAX TR and SHR MAX AX. The former is the maximum transverse spacing to height ratio of the luminaires when they are side-by-side and the latter is the maximum axial spacing when they are end-on and in a line. In your exercise these distances will be found by multiplying together the height (H) between the horizontal reference plane and luminaire plane, and the stated maximum SHR given in the photometric tables for the selected luminaires.

Regulation requirements (Ex. 10). In this exercise you will need to refer to the *IEE Wiring Regulations*, *IEE Site Guide* and relevant Guidance Notes. Look for information concerning the following:

> *Bathrooms* – lighting point, switches, circuit disconnection time, socket outlets, portable and fixed heaters, shaver unit, earth bonding.
> *Cable terminations* – removal of insulation, stress on cables, connection of circuit protective conductors, polarity of conductors.
> *Earthing and bonding* – size of cables to use for main equipotential and supplementary bonding, types of earth clamp and labels.
> *Inspection and testing* – connection and identification of conductors, protection of cables, presence of labels, diagrams, etc., testing procedures and completion of forms (WR1 to WR5).

5.3 Laboratory work (Chapter 4)

Balanced and unbalanced (Experiment 4.1). These are terms used to distinguish between the state of a load when it is connected either in star or delta. A balanced state is one in which the phase or line currents all have the same value whereas an unbalanced state means that they have different values. **Note:** Three-phase supplies to premises are mostly four-wire, i.e. they contain a neutral to cover the possibility of an unbalanced system occurring, which of course happens when single-phase loads are connected. When designing an installation, the designer attempts to split the connected loads over the three phases in order to achieve balance conditions. In achieving this, he is able to make a better and more economical selection of switchgear and cable for the installation.

RC in series (Ex. 4.2). This connection results in the circuit taking a leading power factor. By adjusting the resistor, you will see that the potential difference across it decreases and because the capacitor has a fixed value the phase angle increases. You should note the increase in current.

RL in series (Ex. 4.3). This connection produces the opposite effect of the RC series circuit in that a lagging power factor occurs. By reducing the value of resistance in circuit the p.d. across the coil increases, causing the phase angle to increase. You should note the increase in current.

RLC in series (Ex. 4.4). This is an experiment to show how series connected components affect the current and voltage quantities of an a.c. circuit. You will find that the capacitor is the dominant component making the current lead. The inductor (of negligible resistance) attempts to neutralise the capacitor's effect on the circuit. **Note:** If the two reactive components match each other in terms of their ohmic reactance, the electrical energy in both components begins to oscillate between them. This condition occurs at the particular frequency called resonant frequency.

RLC in series/parallel (Ex. 4.5). This is an arrangement showing the current distribution in a parallel circuit. It will be seen that supply current is not the algebraic sum of the branch currents: instead, it is the phasor sum. By varying the capacitor, circuit conditions change from a lagging power factor to a leading power factor.

SOX lamp circuit (Ex. 4.6). This experiment is similar to experiment 4.5 and is a practical example of mixed circuit components. The term *efficacy* should not be confused with *efficiency* since the former expresses the ratio of output to input in different units (lumens/watt). The latter expresses output to input in the same units (e.g. watts/watt). **Note:** In this experiment the control gear losses need to be added to the lamp's wattage in order to determine the luminaire's efficacy.

Shunt and multipliers (Ex. 4.7). These are resistors which are used for diverting current in d.c. measuring instruments in order to extend the scale range. Shunts have a low resistance value and are used in ammeter circuits, overload protection and control devices. Multipliers have a high resistance and are mostly found in voltmeter circuits.

Frequency variation (Ex. 4.8). The frequency generated by the public supply utility is 50 Hz but any variation of this will affect circuit components possessing reactance. Frequency is proportional to inductive reactance (X_L) and inversely proportional to capacitive reactance (X_C).

Universal motor (Ex. 4.9). This is a series wound motor which can be run on a d.c. or a.c. supply. Its speed characteristics are such that it will race to a high speed if it is left lightly loaded. The experiment is to find its speed–torque characteristics. In your report, you should state a number of applications for this motor. For further information see page 54–56, chapter 3 of the author's *Part 2 Studies: Science* book.

Volt drop in a cable (Ex. 4.10). See the above comments for Task 2.4 in this chapter.

Transformation ratios (Ex. 4.11). A transformer is a piece of electrical equipment designed to change the voltage and current levels. The change is brought about by the number of turns on its primary and secondary windings. This experiment attempts to verify the voltage, turns and current ratios.

Temperature coefficient of resistance (Ex. 4.12). This is a term which relates to the fractional increase per degree Celsius of resistance at 0°C. Most metal conductors (like tungsten) increase in resistance when the temperature increases and they are said to have positive temperature coefficients. Carbon, however, has a negative temperature coefficient. The graphs you construct in this experiment will show these two conditions.

Cage induction motor (Ex. 4.13). The operation of this motor is explained in the author's *Part 2 Studies: Science* book. Very briefly, it is normally a constant speed motor with its rotor chasing after the synchronous speed developed in its stator. The rotor cannot catch up the stator field and it settles down to a speed below synchronous speed. The term 'slip' is used to describe the difference between the two speeds. The experiment investigates the efficiency of the motor using a brake test which enables the motor's output power to be found for various load conditions.

Power factor corrections (Ex. 4.14). Several of the experiments mentioned above are associated with finding the circuit power factor. In this practical experiment a bank of capacitors is used to inject the leading component. Whilst the capacitors will not affect the light output from the fluorescent lamp, they are used to provide minimum current conditions for the circuit.

Rectification (Ex. 4.15). This is a process of converting alternating current into direct current (often described as unidirectional current). In this experiment you are to investigate the waveforms of rectifier elements to give half-wave and full-wave rectification. You should notice a difference in the lamp's brightness when full-wave rectification is used.

Audio frequency amplifier (Ex. 4.16). See Task 2.17 and comments in this chapter. If set up as indicated, you should obtain an output of 2 V and a voltage gain of 200. With an input of 50 mV the negative output cycle will be distorted.

Multimeter (Ex. 4.17). A meter that is capable of reading a range of electrical quantities. The one used in this exercise is an analogue bench type instrument used in college laboratories.

Star-delta starter (Ex. 4.18). One of several methods of starting a three-phase induction motor by reducing the starting voltage to 58%. By doing this the inrush current is also reduced.

Earth fault loop impedance (Ex. 4.19). This is the impedance of the path taken by an earth fault current as it passes from the point of fault back to the supply transformer and then returns to the point of fault again. It is intended that this path will possess a low ohmic value to enable protective devices to operate and safely disconnect the circuit.

Transformer tests (Ex. 4.20). One method of finding the efficiency of a transformer is to make tests on it. An open circuit test finds the transformer's core (or iron) losses which are its no-load losses. The closed circuit test finds the transformer's winding (or copper) losses since its rated current is passed through the primary and secondary windings.

5.4 References

In order for you to extend your knowledge on topics in this book, you should make reference to *Part 2 Studies: Science* and *Part 2 Studies: Theory* of this series.

Health and Safety at Work Act, 1974, HMSO
SI 1989/2209 *The Construction (Head Protection) Regulations*, 1989, HMSO
SI 1988/1057 *The Electricity Supply Regulations*, 1988, HMSO
SI 1989/635 *Electricity at Work Regulations*, 1989, HMSO
Institution of Electrical Engineers Regulations for electrical installations: sixteenth edition, 1991, Stevenage, Herts

The following guidance notes are obtainable from the Health and Safety Executive, HMSO:

GS 6 *Avoidance of danger from overhead electrical lines*, 1991
GS 24 *Electricity on construction sites*, 1983
GS 27 *Protection against electric shock*, 1984
GS 31 *Safe use of ladders and trestles*, 1984
GS 33 *Avoiding danger from buried electricity cables*, 1985
GS 34 *Electrical safety in departments of electrical engineering*, 1986
GS 37 *Flexible leads, plugs, sockets, etc*, 1985
GS 38 *Electrical test equipment for use by electricians*, 1991
PM 32 *Safe use of portable electrical apparatus (electrical safety)*, 1990
PM 38 *Selection and use of electric handlamps*, 1992
HS(G) 13 *Electrical testing: safety in electrical testing*, 1980
HS(G) 38 *Lighting at work*, 1987
HS(R) 25 *Memorandum of Guidance on the Electricity at Work Regulations*, 1989
HS(R) 23 *Guide to the Reporting of Injuries, Diseases and Dangerous Occurrences Regulations, 1985*, 1986